Achieving QTS

Meeting the Teachers' Standards Framework

Primary Science

Knowledge and Understanding

Sixth edition

Primary Science

Knowledge and Understanding

Sixth edition

Graham Peacock • John Sharp
Rob Johnsey • Debbie Wright

 |

Los Angeles | London | New Delhi
Singapore | Washington DC

Learning Matters
An imprint of SAGE Publications Ltd
1 Oliver's Yard
55 City Road
London EC1Y 1SP

SAGE Publications Inc.
2455 Teller Road
Thousand Oaks, California 91320

SAGE Publications India Pvt Ltd
B 1/I 1 Mohan Cooperative Industrial Area
Mathura Road
New Delhi 110 044

SAGE Publications Asia-Pacific Pte Ltd
3 Church Street
#10-04 Samsung Hub
Singapore 049483

Editor: Amy Thornton
Production controller: Chris Marke
Project management: Deer Park Productions
Marketing manager: Catherine Slinn
Cover design: Toucan Design
Typeset by: PDQ Typesetting Ltd
Printed by: MPG Books Group, Bodmin, Cornwall

© 2012 Graham Peacock, John Sharp, Rob Johnsey,
Debbie Wright

First published in 2000 by Learning Matters Ltd

Reprinted in 2001(twice). Second edition published
in 2002. Reprinted in 2003 (twice), 2004 (twice),
2005 and 2006 (twice). Third edition published in
2007. Reprinted in 2007 and 2008. Fourth edition
published in 2009. Reprinted in 2010. Fifth edition
published in 2011. Reprinted in 2011 (twice).
Sixth edition published in 2012.

Library of Congress Control Number: 2012934558

British Library Cataloguing in Publication data

A catalogue record for this book is available from the
British Library

ISBN 978 0 85725 899 1

ISBN 978 1 44625 688 6 (hbk)

FSC

Contents

1
Introduction

About this book

This book has been written to support the needs of all primary trainees on all courses of initial teacher training in England and other parts of the UK where a secure subject knowledge of science is required for the award of Qualified Teacher Status (QTS) or its equivalent. This book will also be found useful by Newly Qualified Teachers (NQTs), mentors, curriculum co-ordinators and other professionals working in education who have identified aspects of their science subject knowledge which require attention or who need a single resource to recommend to colleagues.

At the time of writing this book, as at various other times in education, there are many possible changes to the curriculum and to the requirements for the subjects within it. The government that took office in May 2010 was planning a reform of Early Years policy that will affect the Early Years Foundation Stage (EYFS). The Department for Education (DfE) was undertaking a review of the primary National Curriculum. This book includes information on the statutory programmes of study for National Curriculum science, which maintained schools must follow until the new National Curriculum is in place in 2014, and on the Early Learning Goals for children in the Early Years. In any transitional period, you will need to understand about what curriculum requirements were in place before the new arrangements and teachers you work with may retain in their practice elements of the earlier ways of working. You will certainly hear colleagues discussing the differences between curriculum initiatives and referring to former frameworks, for example the Primary National Strategy for Teaching Literacy and Mathematics. Schools may also still be using units of the exemplar scheme of work for science produced by the Qualifications and Curriculum Authority (QCA). More details about this are provided in the section on curriculum context later in this introductory chapter.

Features of the main chapters of this book include:

- clear links with the Teachers' Standards;
- information about the curriculum context, including the science National Curriculum for England and the Early Years Foundation Stage;
- science knowledge and understanding;
- reflective and practical activities for you to undertake, many of which are related to pedagogy;
- research summaries that give additional background insights into how children's understanding of science develops;
- a summary of key learning points;
- details of publications referenced in the chapter;
- suggestions for further reading on the aspect in question.

There are also self-assessment questions so that you can check on how well you have assimilated the knowledge and understanding. The answers to these questions are contained in a separate chapter. The book also contains a glossary of terms.

> NOTE: Throughout this book, technical terms are often used before they are fully discussed in a later section. We therefore suggest that when you come across a word in **bold** (indicating a glossary reference) you look up its meaning in the glossary before continuing with your reading.

Science subject knowledge really does matter!

A secure subject knowledge of science is widely acknowledged as a critical factor at every point in the complex process of teaching science itself. Few nowadays would argue that planning, teaching and assessing science lessons, setting learning outcomes, choosing appropriate activities and resources, identifying children's errors and misconceptions, asking and handling questions could be achieved without knowing some science in the first place. Some research goes so far as to suggest that a lack of science subject knowledge combined with a lack of confidence in how to teach science may have a severely limiting effect on children's learning. Knowing about science (science content knowledge) and knowing how to teach science (science pedagogical knowledge) are inextricably linked. One of the keys to classroom success is making these connections in science teaching. Readers interested in finding out more about the role of science subject knowledge in teaching are directed towards the references in the Further reading section at the end of this chapter.

The Teachers' Standards for Primary Science

The Teachers' Standards in England (DfE, 2011a) came into force from 1 September 2012, replacing the standards for Qualified Teacher Status (QTS), the Core Professional Standards and the General Teaching Council for England's *Code of Conduct and Practice for Registered Teachers*. These standards define the minimum level of practice expected of all teachers from the point of being awarded QTS. This book refers mostly to the science-related subject standards you will be required to demonstrate in order to be awarded QTS. (See Sharp et al., 2012 for pedagogical and professional theory and practice.)

By reading all of the chapters in this book and successfully completing the practical tasks within them, trainees will have begun to address the standards.

Curriculum context

As referred to at the start of this chapter, the curriculum is likely to change, and this may very well happen several times over your career as a teacher. Because of this, we have focused in this book on the core areas of science subject knowledge and understanding that you will need to develop the science knowledge and understanding and investigative skills of the children you work with, so that you can use this to underpin your planning for whatever curriculum is in place. However, you will need to know about the context of the curriculum prior to any new initiatives so that you can understand how your children have been taught and why. Therefore, we have included context about the primary National Curriculum programmes of study and the Early Learning Goals for children in the Early Years Foundation Stage.

Science in the National Curriculum

Science in the National Curriculum (DfEE/QCA, 1999) is organised on the basis of four key stages. Key Stage 1 for 5- to 7-year olds (Years 1 and 2) and Key Stage 2 for 7- to 11-year olds (Years 3 to 6) are for the primary key stages. The components of each key stage include programmes of study, which set out the science that children should be taught. The programmes of study are referenced e.g. Sc2/2a. At Key Stage 2 this refers to the functions and care of teeth.

Attainment targets set out the science that children should know and be able to do, and level descriptions describe what children working at a particular level should be able to demonstrate. Science in the National Curriculum is a minimum statutory requirement. Since its introduction in 1989, it has been significantly revised three times. The programmes of study include requirements for Scientific enquiry (ideas and evidence in science; investigative skills); Life processes and living things (life processes; humans and other animals; green plants; variation and classification; living things in their environment); Materials and their properties (grouping materials; changing materials; separating mixtures of materials – Key Stage 2 only); and Physical processes (electricity; forces and motion; light and sound; the Earth and beyond – Key Stage 2 only).

As this edition went to print, the DfE were reviewing the 1999 National Curriculum. *The Framework for the National Curriculum*, a report by the expert panel for the National Curriculum Review (DfE, 2011a), recommended that the structure of the Key Stages changed to Key Stage 1 for 5–7 year olds (Years 1 and 2), Lower Key Stage 2 for 7–9 year olds (Years 3 and 4), and Upper Key Stage 2 for 9–11 year olds (Years 5 and 6).

The report reaffirmed that the National Curriculum is a minimum statutory entitlement for children and that schools have the flexibility and freedom to design a wider school curriculum to meet the needs of their pupils and decide how to teach it most effectively. The report recommended that the new National Curriculum should retain discrete and focused elements within the Programme of Study for science.

The Bew Review (Bew, 2011) into Key Stage 2 testing, assessment and accountability identified that that teacher assessment is the most appropriate form of assessment for science at the end of Key Stage 2. In their response to the review (DfE, 2011b), the Government accepted this view and so pupil and school level data will continue to be based on teacher assessment judgements. The Government feel that it is important that national performance in science should continue to be monitored and so the Standards and Testing Agency will continue to develop and administer national sample tests in science.

Furthermore, subject to the outcomes of the National Curriculum Review, the Government plan to ask the Standards and Testing Agency to develop a system of pupil-level science sampling in order to provide much greater detail about the attainment of pupils nationally across the whole science curriculum.

Use of an exemplar scheme of work for science at Key Stages 1 and 2 (QCA/DfEE, 1998, with amendments 2000) was entirely optional. Designed to help implement science in the National Curriculum, many schools used and adapted it for their own needs. The scheme was presented as a series of units that attempted to provide continuity and progression in primary science provision between Years 1 and 6. Guidance was offered on:

- the nature and place of each unit;
- how each unit builds on previous units;
- technical scientific vocabulary;
- resources;
- expectations;
- teaching activities;
- teaching outcomes;
- health and safety;
- ICT links.

You may find that schools and individual teachers you work with are still using whole units, adapted units or elements of this scheme of work in their medium-term and long-term planning for science.

Early Years Foundation Stage

The Early Learning Goals (DfEE/QCA, 1999) describe what most children should achieve by the end of their Reception year. Together with the *Statutory Framework for the Early Years Foundation Stage* (DCSF, 2008), they identify features of good practice during the foundation stage, which begins when children reach the age of three, and set out the Early Learning Goals in the context of six areas of learning. At the time of writing this edition the Tickell Review, *The Early Years: Foundations for Life, Health and Learning* (Tickell, 2011), had recommended a number of changes for EYFS reforms. For example, it has recommended that the EYFS reform changes the six areas of learning in development into three prime areas and four specific areas in which the prime areas are applied:

Prime areas:
- personal, social and emotional development;
- communication and language;
- physical development.

Specific areas:
- literacy;
- mathematics;
- expressive arts and design;
- understanding the world.

Although science is most likened to the previous area of learning and development, Knowledge and Understanding of the World, in the new EYFS Framework it will be necessary to focus on the prime areas and apply the knowledge, skills and understanding that young children are developing on them within a context of science through the specific area Understanding the World.

Outcomes

By using this book to support your own subject knowledge development, you will be able to learn the knowledge and develop the understanding that you require to teach primary science and the appropriate elements of Knowledge and Understanding of the World.

So that you can check on how well you have assimilated the subject knowledge and test your understanding, you may wish to try the self-assessment questions related to each aspect that we address. You will find these in a separate section towards the end of the book. The answers to these questions are provided for you in a separate chapter.

For those undertaking credits for a Masters Degree, we have included suggestions for further work and extended study at the end of each chapter in a section called 'M-Level Extension'.

REFERENCES REFERENCES **REFERENCES** REFERENCES REFERENCES

To support you in understanding the curriculum context, you may find it helpful to refer to some of the following documentation:

Bew, P. (2011) *Independent Review of Key Stage 2 Testing, Assessment and Accountability: Final Report*. London: DfE.

DCSF (2008) *Statutory Framework for the Early Years Foundation Stage*. Nottingham: DCSF.

DfE (2011a) *The Framework for the National Curriculum. A Report by the Expert Panel for the National Curriculum Review*. London: DfE.

DfE (2011b) *Independent Review of Key Stage 2 Testing, Assessment and Accountability: Government Response*. London: DfE.

DfEE/QCA (1999) *Science: the National Curriculum for England*. London: HMSO.

QCA/DfEE (1998, with amendments 2000) *Science: a Scheme of Work for Key Stages 1 and 2*. London: QCA.

Sharp, J., Peacock, G., Johnsey, R., Simon, S. and Smith, R. (2012) *Primary Science: Teaching Theory and Practice*. Exeter: Learning Matters.

Tickell, C. (2011) *The Early Years: Foundations for Life, Health and Learning. An Independent Report on the Early Years Foundation Stage to Her Majesty's Government*. London: DfE.

FURTHER READING FURTHER READING **FURTHER READING** FURTHER READING

Readers interested in finding out more about science subject knowledge and teaching are directed towards the following sources.

Byrne, J. and Sharp, J. (2006) Children's ideas about micro-organisms, *School Science Review*, 88(322), 71–79.

DfE (2011) *Teachers' Standards*. Available at www.education.gov.uk/publications.

Harlen, W. and Qualter, A. (2004) *The Teaching of Science in Primary Schools*. London: David Fulton.

Howe, A., Davies, D., McMahon, K. Towler, L. and Scott, T. (2005) *Science 5–11: A guide for teachers*. London: David Fulton.

Office for Standards in Education (2011) *Successful Science*. London: Ofsted.

Sharp, J., Peacock, G., Johnsey, R., Simon, S. and Smith, R. (2012) *Primary Science: Teaching Theory and Practice*. Exeter: Learning Matters.

Summers, M., Kruger, C., Childs, A. and Mant, J. (2001) Understanding the science of environmental issues: development of a subject knowledge guide for primary teacher education, *International Journal of Science Education*, 23(1), 33–35.

2
Functioning of organisms: green plants

Curriculum context

National Curriculum programmes of study

At Key Stage 1, children should be taught about the basic conditions necessary for plant growth, to recognise and name the simple parts of flowering plants, and that flowering plants develop and grow from seeds.

At Key Stage 2, children should be taught about plants and the factors affecting plant growth, nutrition and reproduction in detail. This should include the role of leaves, roots, stems and flowers. Some attention should be given to the main parts of a flower and to the life cycle of a flowering plant.

Early Years Foundation Stage

Children must be supported in developing the knowledge, skills and understanding that help them to make sense of the world. Their learning must be supported through offering opportunities for them to encounter plants in their natural environments and in real-life situations and undertake practical 'experiments'. By the end of the EYFS, children should:

- investigate objects and materials by using all of their senses as appropriate;
- find out about, and identify, some features of living things they observe;
- look closely at similarities, differences, patterns and change;
- ask questions about why things happen and how things work.

Introduction

Green plants grow on every continent and colonise virtually every type of habitat. The diversity of green plants on the surface of the Earth today is quite simply staggering. The importance of green plants, together with other photosynthetic organisms, cannot be overstated. They are the starting points for most food chains and they produce the oxygen which keeps most other living things alive. Studying green plants in the school grounds or within the classroom provides excellent opportunities for children of all ages to develop their scientific skills and to raise and test their own investigative questions. Studying green plants also allows children to participate in major socio-scientific debate (e.g. the use of pesticides, the destruction of the rain forests, and genetically modified plants and plant products).

For more information on plants in general, see Chapter 5 on ecosystems.

Green plants and the plant kingdom

Modern classifications of living organisms within the plant kingdom recognise six main phyla (see Figure 2.1). These have changed over time and vary from source to source but currently include:

- mosses and liverworts (Bryophyta);
- horsetails (Sphenophyta);
- club mosses (Lycopodophyta);
- ferns (Filicinophyta);
- conifers (Coniferophyta);
- flowering plants (Angiospermophyta).

The plant kingdom contains a particularly diverse group of complex, multicellular, eukaryotic organisms, almost all of which live on land. Both sexual and asexual reproduction are common. Green plants are **autotrophs** and capable of making their own food by photosynthesis. The majority of green plants commonly encountered are conifers and flowering plants. **Fungi**, **algae** and **lichens** are no longer considered plants and do not usually appear in modern plant classification systems.

RESEARCH SUMMARY RESEARCH SUMMARY RESEARCH SUMMARY **RESEARCH SUMMARY**

Research into children's ideas about plants from Nursery to Year 6 has been made available by Heath (2008), Bianchi (2000), Jewell (2002), Briten (2006) and Fernandes *et al*. (2006). Findings showed that many of the children involved appeared to grow up with a limited knowledge and understanding of what plants are, the factors responsible for their growth, how they reproduce, and the diversity within the plant kingdom as a whole. Stereotypical images of plants as simply 'flowers', and fruits as apples and oranges, quite often persisted until well beyond the primary years. Children have a restricted idea about the term 'seed', as reported by Jewell (2002a,b). Many children found difficulty considering plants to be alive because of a lack of obvious movements and noises. Many children relied heavily on restricted personal experiences rather than on knowledge gained to explain phenomena associated with plants and some often became confused as their intuitive ideas about plants were challenged.

The use of scientific vocabulary in children's explanations was found to improve with age but the mixed scientific and everyday use of common terms associated with plants (e.g. seeds, fruits, vegetables and weeds) remained problematic. The concept of 'plant' is clearly one that requires some attention and careful development.

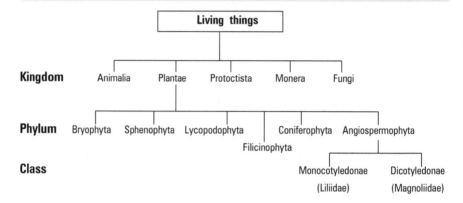

Figure 2.1 Partial classification of green plants

Mosses and liverworts
Bryophytes, including common mosses and liverworts, are among the most 'primitive' of all green plants. Most are relatively small and possess only simple root-like rhizoids, stems and leaves. Bryophytes are also non-vascular – the food and water conducting phloem and xylem vessels found in other green plants are absent. Most bryophytes live where it is permanently wet or damp. The sphagnum mosses found in peat bogs and other waterlogged soils are perhaps the best-known examples.

Horsetails, club mosses and ferns
Sphenophytes, lycopodophytes and filicinophytes, including horsetails, club mosses and ferns, are more advanced than the bryophytes. A simple vascular system of phloem and xylem vessels for conducting food and water around each plant and for support is present. Like bryophytes, the members of these groups live in permanently wet or damp conditions. Sphenophytes, lycopodophytes and filicinophytes are often referred to collectively as pterydophytes.

Conifers
Coniferophytes are essentially cone-bearing trees with needle-like leaves. Examples include varieties of pine, spruce and cedar. All possess a highly developed vascular system of phloem and xylem vessels. Many reach great heights, live for long periods of time and can be found growing in a wide range of conditions. The California redwood tree (*Sequoia sempervirens*) is one of the tallest plants known, growing to heights of well over 100 metres. Coniferophytes are mostly 'evergreen' and retain their leaves throughout the year. One notable exception is the larch.

Coniferophytes and other groups of plants including ginkgos (maidenhair trees) and cycads (tropical palm-like trees) are often referred to collectively as gymnosperms.

Flowering plants

Together with coniferophytes, the angiospermophytes or flowering plants are among the most advanced plants known. They are also the most dominant. All possess a highly developed vascular system of phloem and xylem vessels and exhibit other forms of specialisation. Like coniferophytes, they can be found growing in a wide range of conditions. Angiospermophytes can be divided into two groups or classes, the Monocotyledonae or monocots (more recently Liliidae) with one cotyledon or specialised leaf in their seeds (includes lillies, irises, daffodils and grasses) and the Dicotyledonae or dicots (more recently Magnoliidae) with two cotyledons or specialised leaves in their seeds (includes most flowering trees, shrubs and herbaceous or non-woody plants).

Life processes

In common with all other living organisms, green plants display certain characteristics which demonstrate that they are alive and carry out certain processes in order to stay alive. These include:

- movement;
- respiration;
- sensitivity;
- growth;
- reproduction;
- excretion;
- nutrition.

Movement

Movement in plants is not immediately obvious. Most plants are firmly rooted to the ground and certainly not capable of locomotion. Movement does take place, of course, but usually slowly and over an extended period of time. There are two main types of plant movement:

1. tropic movements (directional movements associated with growth and in response to an external stimulus);
2. nastic movements (non-directional movements in response to an external stimulus and not necessarily associated with growth).

The most commonly observed tropic movements occur as stems and leaves grow towards light (phototropism), as roots grow into soil in the direction of gravity (positive geotropism or gravitropism) and as shoots grow upwards against gravity (negative geotropism or gravitropism). Movement also occurs as lateral roots extend outwards from a plant's main root (diageotropism). Certain climbing plants like ivy produce tendrils which grow around the objects they touch (thigmotropism). The most commonly observed nastic movements occur as the petals and florets of flowers like daisies and crocuses open and close at certain times of the day and night (photoperiodism). Nastic movements are most spectacular in the modified leaves of the Venus fly-trap (*Dionaea muscipula*), which close around insects, and the leaves and branches of the sensitive plant (*Mimosa pudica*), which sag and droop when shaken.

For more information on respiration, see the section on energy transfers in biological processes in Chapter 9.

Respiration

Plants need energy in order to move, to grow, to carry out the reactions or metabolic processes which take place within **cells**, to transport substances to where they are needed, and for the active uptake of minerals from the soil. This energy is made available in all living plant cells during respiration. Cellular respiration in plants essentially involves a reaction between simple food substances like glucose and oxygen. Glucose and other simple sugars are manufactured in the leaves and other photosynthetic tissues of plants by **photosynthesis**. Photosynthesis also requires energy but this is obtained from sunlight and not from respiration. The oxygen required for respiration enters plants mostly through their leaves. Cellular respiration in plants is most simply represented as follows:

$$\text{glucose } + \text{ oxygen } \rightarrow \text{ carbon dioxide } + \text{ water } + \text{ energy}$$

To reach the parts of plants where energy is needed but where photosynthesis does not take place, in roots for example, simple sugars are transported around plants in phloem. Roots receive some of their oxygen directly from the air spaces in soil and water.

Sensitivity

Plants do not have a nervous system. Nevertheless, plants do respond to certain stimuli in the same way that most other living organisms do. This is usually made possible within plants by certain chemical substances which are activated in a number of different ways. Plants are most obviously stimulated by light, gravity and touch, when responses to stimuli can be detected in movements. The seeds of some plants are sensitive to seasonal variation too. They respond by lying dormant until conditions are best suited for germination. Being sensitive, therefore, allows seeds to germinate and develop into healthy plants, brings leaves into the best orientation for efficient photosynthesis, allows roots to anchor plants and to extend outwards to obtain water and minerals from the soil, and allows plants to climb to take advantage of light and other conditions and to flower at a time when certain insect species are most available for pollination.

Growth

Plants may grow throughout their entire lives, changing constantly in both size and mass. However, plants do not grow all over. Plant growth takes place at special growth sites at the tips of roots and shoots, where plants bud along stems, and between the phloem and xylem vessels of a plant's vascular system. During plant growth, great numbers of new cells are produced which absorb water and become enlarged or elongated. Plant growth therefore results from the division and expansion of cells. As plants grow taller they also need to be supported. Support may come from the ribs of stems, the pipe-like phloem and xylem vessels of vascular systems, the additional lignification of xylem, and the extension of roots. In trees, support comes from wood, the fully lignified xylem of old growth. Turgor pressure, the fluid pressure from within cell vacuoles, also helps. Plants not watered regularly wilt when cells become flaccid (no longer turgid). The energy required for growth and the manufacture of new cellular material comes from respiration. The nature, direction and rate of plant growth is controlled by the production of chemical substances like auxins and gibberellins, hormone-like, plant growth substances which promote or inhibit cell division. The same substances are also associated with some plant movements and plant sensitivity.

Reproduction

Plant reproduction ensures the continuity of the species. Plants are capable of both sexual and asexual reproduction and reproduction strategies can be complex and varied. Bryophytes, for example, are capable of sexual reproduction only when it is damp or permanently wet. Male and female sex **organs** are located on separate plants. The male gametes or sex cells (sperms) have to 'swim' using flagella in order to reach the female gametes or sex cells (eggs). Following sexual reproduction, bryophytes pass through a spore-forming or asexually reproducing stage with each spore potentially giving rise to a new fully developed plant. Coniferophytes, for example, produce both male and female cones. Pollen grains, from which male gametes or sex cells develop, are transferred to female cones by the wind. Following fertilisation it may take years for the seeds produced to develop and ripen ready for dispersal. Angiospermophytes have flowers for the purpose of sexual reproduction. This will be dealt with in more detail later in this chapter. Angiospermophytes are also capable of reproducing asexually (vegetative propagation). Mechanisms involve the production of aerial runners or stolons (e.g. strawberries), underground stems or rhizomes (e.g. grasses) and bulbs, corms and tubers (e.g. daffodils, crocuses and potatoes).

Excretion

Excretion in plants involves the removal of unwanted waste or by-products associated with plant cells as they go about performing the metabolic processes that keep them alive. This mainly involves the gases produced as a result of photosynthesis and cellular respiration. During photosynthesis, simple sugars and oxygen are produced from a reaction between carbon dioxide and water using the energy in sunlight. During respiration, carbon dioxide and water are produced from a reaction between simple sugars and oxygen. As the rate of photosynthesis usually exceeds the rate of respiration during the day, plants excrete oxygen. At night, when there is no photosynthesis going on at all, plants excrete carbon dioxide. At dawn and at dusk, photosynthesis and respiration occur at about the same rate. During these times, the oxygen and carbon dioxide produced during both reactions are simply exchanged. Plants do not require the complex respiratory systems found in other living organisms. Leaves, for example, are thin and the distances gases have to travel are small. Oxygen and carbon dioxide are able to enter and exit through stomata or 'pores' by diffusion. In the bark of woody plants like trees, specialised cellular structures called lenticels allow the diffusion and excretion of respiratory gases to take place.

Nutrition

While humans and most other animals must move around for their food and eat what they find (**heterotrophic** nutrition), green plants essentially make their own (**autotrophic** nutrition). The key process in plant nutrition is photosynthesis and the production of simple sugars like glucose from the carbon dioxide in air and the water in soil using the energy in sunlight. If respiration is the process by which the energy in food substances is released and made available to do other things, then simple sugars like glucose constitute the food of plants. Photosynthesis in plants is most easily represented as follows:

For more information on photosynthesis, see the section on energy transfers in biologocial processes in Chapter 9.

$$carbon\ dioxide\ +\ water\ \xrightarrow[\text{(chlorophyll)}]{\text{(energy in sunlight)}}\ glucose\ +\ oxygen$$

Chlorophyll absorbs the energy in sunlight making it available to drive the reaction between carbon dioxide and water. From the simple sugars produced during photosynthesis, plants are able to manufacture carbohydrates (e.g. cellulose and starch), lipids (e.g. oils and waxes) and proteins. The formation of proteins requires certain elements like nitrogen which are also obtained from soil. It is important to note that plants do not take food from the soil or from the air around them. They make their own food in their leaves and other photosynthetic **tissues** by photo-synthesis. Products labelled as 'plant foods' use the term food in a figurative rather than scientific way.

REFLECTIVE TASK

Children's ideas about green plants, from classification to life processes to function, are fascinating. You can explore these ideas and find out more by consulting the published sources provided in the further reading and reference list at the end of this chapter or you might be able to elicit them for yourself while on placement actually teaching science. Knowing what children might think about green plants, what resources and activities would you select to challenge their alternative ideas most effectively? Plant nutrition, for example, is a particularly complex area of science, yet intro-duced at Key Stage 1. How might you introduce the concept of plant nutrition, or any other plant concept for that matter, without adding to their confusion?

Cells, tissues and organs: levels of organisation

Plants are made from cells. Plant cells, like the cells in human bodies, are highly specialised and carry out particular functions within a plant (e.g. root hair cells, guard cells, phloem cells, xylem cells and palisade cells). This division of labour is reflected in different levels of organisation.

- A group of specialised plant cells of the same type that work together to carry out a particular function is called a tissue (e.g. the epidermis, the palisade mesophyll, the spongy mesophyll and the phloem and xylem of leaves).
- A group of tissues that work together to carry out a particular function is called an organ (e.g. root, stem, leaf and flower).

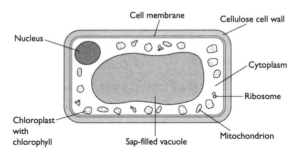

Figure 2.2 Plant cell – green tissue (actual size 0.01 mm across)

Many plant cells have several features in common (see Figure 2.2).

- A cell wall.
- A cell membrane.
- Cytoplasm.
- A single nucleus and other organelles or 'mini-organs', including mitochondria and ribosomes.
- Chloroplasts (specialised organelles referred to as plastids – green tissue cells).
- A vacuole.

For more information on cells in general, see Chapter 4 on continuity and change.

The outermost part of a plant cell, the cell wall, is made from non-living layers of cellulose fibres which provide some support and strength. The cell walls allow all liquids and dissolved substances to pass through. They are not selectively permeable like the cell membrane which encloses the contents of the cell. Cytoplasm is a jelly-like substance mostly made from water. Cytoplasm contains all a plant cell's organelles or 'mini-organs'. Plant cells are **eukaryotic**. The nucleus contains a plant cell's genetic material or plant DNA which ultimately determines what type of cell it is and controls what it does. The genetic material or DNA also has the ability to reproduce itself in a process known as replication. This is important during cell division as plants grow. Other important organelles or 'mini-organs' include mitochondria, where cellular respiration takes place, and ribosomes, where proteins are made. Some cells may also contain granules of starch as energy stores. Chloroplasts are specialised organelles referred to as plastids. They contain the green pigment **chlorophyll**, an important substance responsible for bringing about photosynthesis. Chloroplasts are usually found in the green tissue cells of leaves and stems. Roots and many flowers are not green, do not contain chloroplasts and do not photosynthesise. Vacuoles are large, usually permanent, fluid or sap-filled cavities which exert an outward pressure on a cell, helping to maintain its shape and provide support and strength. Sap is a watery mixture of various substances including salts, pigments and waste products. The vacuoles of some plants often contain plant toxins, which act as deterrents to herbivores.

The structure of flowering plants

Flowering plants, the most dominant forms of vegetation on Earth, vary enormously in size, shape, colour and form. While there is no such thing as a typical flowering plant, they do possess many features in common (see Figure 2.3). These include:

- a root system, usually beneath the ground;
- a shoot system of stems, leaves and flowers, usually above the ground.

Roots
Roots usually anchor plants firmly in one place and prevent them from falling over. Roots are also responsible for the uptake of water and dissolved minerals from the soil. This is often made possible by specialised root hair cells which can occur over an entire root system. Water and dissolved minerals entering the roots pass into the xylem of a plant's vascular system. From there, the water and dissolved minerals are eventually drawn upwards through the stem to the leaves and flowers. The upward passage of water drawn from the soil is known as transpiration. The most important root of some plants is the main or tap root. In such cases, all other roots

emanate from this structure. Plants without obvious tap roots usually have fibrous roots.

Stems

Stems usually hold plants upright, spread out leaves for photosynthesis and elevate flowers for pollination. They also contain a plant's vascular system of phloem and xylem vessels. In dicot flowering plants, the phloem and xylem cluster in vascular bundles arranged in a ring. The plant cells within the ring make up the pith, while the plant cells between the ring and the outermost 'skin' or epidermis make up the cortex. Both pith and cortex provide 'packing' which help with support. In monocot flowering plants, the phloem and xylem are more irregularly distributed. With some plants, lateral or axillary buds give rise to branches with leaves and occasionally flowers. Terminal buds at the tips of shoots allow for the upward growth of stems. These may also produce flowers. Regions of stems where leaves and buds arise are called nodes. Where present, small leaves at the bases of flower stalks are called bracts and outgrowths from the bases of leaves are called stipules.

Leaves

The leaves of most plants are usually attached to stems by leaf stalks or petioles. In dicots, the petioles continue into a 'midrib' or 'main vein' from which a network of smaller 'veins' and 'veinlets' emerges. In monocots, the 'veins' and 'veinlets' appear parallel. The 'veins' and 'veinlets' of leaves contain the vascular bundles of phloem and xylem vessels. Xylem conducts water and dissolved minerals from roots to leaves. Phloem usually transports simple sugars and other substances away from leaves to stems, flowers and roots. The principle function of leaves is to manufacture food by photosynthesis. The leaves in some plants including gorse have been reduced to spines. The arrangement and shape of leaves can be important in plant identification. While most leaves are green due to the presence of chlorophyll, others do vary. Trees which begin to lose their leaves in winter reveal colours produced by pigments usually masked by chlorophyll. These include oranges (carotene), yellows (xanthophyll), reds and purples (anthocyanin) and browns (tannin).

Flowers

Flowers are the structures of plants responsible for reproduction and seed formation. Single flowers may be carried on single flower stalks or pedicels but more often these are grouped to form an inflorescence. Flower heads usually consist of five main elements.

1. Petals (often brightly coloured and scented – not always present).
2. Thin and papery sepals which protect the flower in bud (not always present).
3. The male reproductive organs or stamen (including the anther and filament).
4. The female reproductive organs or carpel (including the stigma, style and ovary).
5. An expanded end of stalk or axis called the receptacle which supports the flower and other structures (not always present).

Like leaves, flower colour and the number and arrangement of petals can be used to identify plants. Many intricate flower shapes, markings and scents have evolved as a result of the close relationship with particular pollinators (e.g. bees, butterflies).

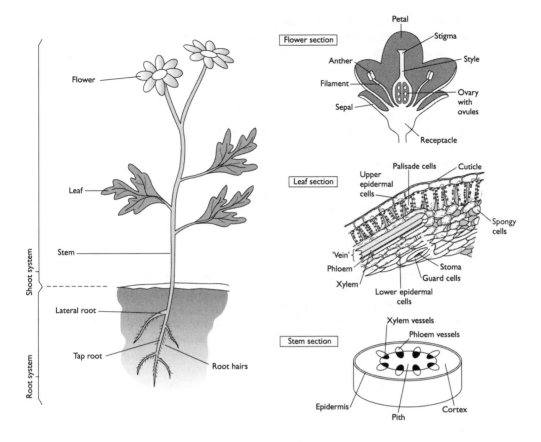

Figure 2.3 Flowering plant

Plants usually flower only at certain times of the year. They are sensitive to seasonal change. This is made possible by a substance called phytochrome.

Photosynthesis

Photosynthesis is the process by which all green plants use the energy in sunlight to convert the carbon dioxide in air and the water in soil into simple sugars like glucose and oxygen. They are able to do this because of a green pigment called chlorophyll. Chlorophyll is an important substance contained within the chloro-plasts of plant cells. Chlorophyll helps certain reactions to proceed without actually taking part in the reactions themselves. In this case, the chlorophyll absorbs the energy in sunlight and makes it available to react the carbon dioxide with the water. Plants use the glucose and other simple sugars they make in a number of different ways.

- As a source of energy.
- To produce sucrose for storage in fruits.
- To produce starch for storage in stems and in seeds.
- To produce cellulose for cell walls, waxes and oils for cell membranes, and proteins for cytoplasm, pigments and enzymes.

The leaves of many plants are large, flat and thin in order to trap sunlight and to make the process of photosynthesis particularly efficient. A cross-section taken through a leaf blade or lamina reveals different layers of mostly specialised cells. These include:

- a waxy cuticle to reduce water loss by evaporation and to protect the upper surface (not always present);
- the upper epidermis, which prevents plant pathogens from entering the leaf and helps maintain leaf shape by secreting a waxy cuticle (if present);
- the palisade mesophyll, where most photosynthesis takes place (palisade cells are filled with chloroplasts);
- the spongy mesophyll, where loosely fitting cells of different shape store photosynthetic products and allow gas movements by diffusion;
- the lower epidermis, with pairs of guard cells surrounding stomata or 'pores' which open during the day and close at night or when water is scarce.

The rate of photosynthesis is limited by several factors, including light intensity (slow on dull days and fast on bright sunny days), temperature (slow when cold or excessively hot as chlorophyll efficiency is reduced), the concentration of carbon dioxide in air (not usually a significant factor) and water availability (when scarce, the stomata close to prevent further water loss thus reducing the rate of photosynthesis).

PRACTICAL TASK PRACTICAL TASK PRACTICAL TASK PRACTICAL TASK

Key Stage 2 children can produce a labelled and annotated drawing of a flowering plant. Key Stage 1 children will focus on the leaves, stems and roots rather than the detail of the flower. Systematically dissect the plant paying particular attention to the flower itself. Can you identify the male and female reproductive organs? Can you find any ovules within the ovary? Take care to ensure that the plant you choose is safe to handle (e.g. doesn't sting or cause allergic reactions, isn't poisonous, and doesn't stain clothing or skin).

Transport systems in flowering plants

The phloem and xylem vessels involved in transporting materials around plants usually run alongside each other in clusters referred to as vascular bundles. The vascular bundles are easily identified as they are located within the 'veins' and 'veinlets' of plants. Vascular bundles provide the link between all parts of a flowering plant.

Phloem

Phloem vessels are long 'tubes' of living cells with perforated end walls or sieve plates to allow sap to flow through them. Phloem cells have no large vacuoles or organelles including a nucleus in order to make this process more efficient. Phloem conducts simple sugars and other substances produced in the leaves of plants to where they are needed or stored (e.g. stem, roots, flowers, seeds, fruits). The mechanism of sap movement within phloem remains uncertain but probably involves the 'loading' and 'unloading' of chemical substances at production and destination sites, and this requires energy.

Xylem

Xylem vessels are long 'tubes' of dead cells connected end to end with no end walls or sieve plates between them. The walls of xylem vessels are often thickened and impregnated with **lignin** which provides support. Xylem conducts water and dissolved minerals from soil to the leaves and other green tissues where photosynthesis takes place. Movement within xylem is brought about by transpiration. Transpiration is the constant flow of water through a plant from roots to leaves. Water lost from the leaves by evaporation is replaced by water drawn into the roots and up the stem. This is made possible by osmosis. Osmosis and water availability throughout a plant also keeps plants turgid (rigid). Plants which wilt due to lack of water are said to be flaccid. The mechanism of mineral uptake in roots remains uncertain but may require energy.

Sexual reproduction in flowering plants

Unlike many living organisms, green plants cannot move about freely to find a mate and reproduce. Instead, sexual reproduction in flowering plants is brought about by specialised organs in which male sex cells or gametes (sperms) fuse with female sex cells or gametes (eggs). This gives rise to a new plant with the combined characteristics of both parent plants. Plants which reproduce sexually may have male and female reproductive organs in the same flower (bisexual or hermaphrodite), in different flowers of the same plant (monoecious) or in different plants (dioecious). Male reproductive organs are referred to as stamens. A stamen includes:

For more information on reproduction in general, see Chapter 4 on continuity and change.

- a filament which supports an anther;
- an anther which produces pollen.

Female reproductive organs are referred to as carpels. A carpel includes:

- a stigma which receives pollen;
- a style which supports the stigma;
- an ovary which contains female sex cells inside one or more ovules.

The reproductive cycle of a flowering plant proceeds in five well-defined stages.

1. Pollination.
2. Fertilisation.
3. Seed formation.
4. Seed dispersal.
5. Germination.

Pollination

Pollination occurs when pollen is transferred from anther to stigma. This is achieved by the wind, by insects or by other pollinators (e.g. birds, small mammals). Wind-pollinated or anemophilous flowers often lack bright colour and produce little scent or nectar. The anthers are commonly located outside the flower head and exposed to the elements. Insect-pollinated or entomophilous flowers are often brightly coloured, scented and produce nectar in abundance. The anthers are commonly located within the flower head. Self-pollination can occur when pollen transfer takes place in the same flower or between different flowers on the same plant. Cross-pollination can occur when pollen transfer takes place between flowers on different plants. Cross-pollination can result in genetic variation.

Fertilisation

Fertilisation occurs when the nuclei of male and female sex cells meet and fuse. This begins when pollen absorbs moisture from the stigma, swells and develops a pollen tube which extends along the style to the ovary and into an ovule. The nucleus of the male sex cell then passes along the pollen tube and fuses with the nucleus of the female sex cell. Each pollen grain can produce only one pollen tube. The fusing of nuclei in other ovules requires other pollen. At the point of fertilisation, a zygote is formed which then grows and develops into an embryo plant within a seed.

Seed formation

After fertilisation, the ovule becomes a seed containing the embryo plant while the ovary usually develops into a fruit. Petals and stamens wither. At this stage, the embryo plant consists of a tiny root or radicle and a tiny shoot or plumule. The rest of the seed consists of specialised leaves or cotyledons and a hard outer coat or testa. In monocot plants there is only one cotyledon. In dicot plants there are two. Not all fruits are fleshy and edible. Some fruits form hard pods and capsules while others form leathery achenes. Biologically, tomatoes and cucumbers are fruits too, though in everyday language these are more commonly referred to as vegetables.

Seed dispersal

After seeds are formed they must be dispersed if the new plants are to survive (competition for light, nutrients and space). Dispersal mechanisms vary. Some plants produce seeds in pods which are scattered when shaken (e.g. poppy). Some seeds are attached to 'parachutes' or 'wings' and float on the wind (e.g. dandelion, sycamore). Many seeds are eaten by birds and scattered in droppings (e.g. blackberry) or buried and forgotten about by squirrels (e.g. hazelnut). Other seeds are contained within fruits with hooks which attach themselves to passing animals.

Germination

Germination requires water, suitable temperatures and oxygen but not light. In a typical dicot seed, for example, the embryo plant is attached to two cotyledons. Once on the ground, the seed absorbs water and swells. As it does, the testa splits, the radicle emerges and begins to grow down into the soil. The plumule emerges afterwards and grows towards the light. Starch within the cotyledons is used by the embryo as a source of energy for growth. Once anchored in the soil, green leaves open out and begin to photosynthesise (see Figure 2.4).

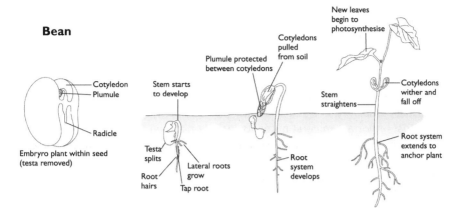

Figure 2.4 Seed germination (may vary from one plant species to another)

PRACTICAL TASK PRACTICAL TASK **PRACTICAL TASK** PRACTICAL TASK

Germinate your own seeds. Almost anything will do but most beans from supermarkets work particularly well. Place a few beans between absorbent paper and the wall of a transparent container like a jar. Keep the paper wet by constantly adding water to the container. The wet paper will keep the beans moist. Do not let the beans become immersed in the water or they will rot. Observe and record what happens before and after germination takes place. Once roots and shoots are fully developed, pot the beans in containers filled with compost or soil. Observe, record and measure the changes which take place as they grow. Represent your findings pictorially and on a suitable graph. Identify your dependent and independent variables. Take care to ensure that the seeds you choose are safe to handle (e.g. do not cause allergic reactions, are not poisonous, and are not treated with any chemical substances).

For Key Stage 1 children, plant a bean seed in a pot. Place it in a cardboard box house with a chimney. The children will be delighted to see the shoot emerge from the chimney as the plant grows towards the light.

Plant health

The health of green plants can be affected in a variety of different ways. Farmers and gardeners are particularly interested in plant health and for obvious, if sometimes different, reasons. Entire food crops can be lost during times of flood or drought with devastating consequences for those dependent on their production. Over-watering, under-watering, providing too much or too little light, draughts and excessive or insufficient room temperatures can produce similar, if not so dramatic, adverse effects. Weeds, plants which grow where they are not wanted, are unsightly and compete with other plants for light, minerals, water and space. Herbicides, insecticides, fungicides and fertilisers are all commonly used in efforts to keep plants healthy and to increase productivity. Selective breeding techniques involving naturally and artificially induced genetic modifications are also used to find disease-resistant or particularly interesting and attractive plant varieties. Common causes of ill health in plants include:

- mineral deficiency;
- invertebrate organisms;
- plant pathogens.

Mineral deficiency

Plants obtain essential minerals from the soil in which they grow. These minerals are usually dissolved in water and enter plants through their roots by absorption. The most important mineral nutrients include nitrogen (N), phosphorus (P) and potassium (K) which are incorporated into cell walls, cell membranes, proteins and plant DNA. Deficiencies result in overall poor growth. Other important minerals include magnesium (Mg), iron (Fe) and sulphur (S) which are used in the manufacture and structure of chlorophyll. Deficiencies result in low light absorption and low sugar production. The availability of all essential minerals can be controlled by soil pH, which affects mineral solubility, and soil texture, including the sand, silt, clay, organic matter content which affects drainage, root penetration and the range of nutrients present.

Invertebrate organisms

Invertebrates, including slugs, snails, earwigs, woodlice and the larval forms of butterflies, moths and other insects, are all capable of doing considerable damage to plants by simply eating them. Roots, leaves and flowers are particularly attractive targets. Aphids and mites are sap-sucking invertebrates. They can remove a plant's sugar-rich solutions directly from phloem and kill plants by infestation.

Plant pathogens

Pathogens include a wide range of fungi, bacteria and viruses. The range of diseases caused by plant pathogens is enormous. Apples that turn brown and rot, for example, have become infected with the fungus *Sclerotina fructigena*. All plant pathogens can be transmitted through air or soil. Bacteria and viruses can also be transmitted by invertebrates and other animals. Bacterial and viral infections are difficult and sometimes impossible to treat. Infected plants are often destroyed.

A SUMMARY OF **KEY POINTS**

> The plant kingdom is divided into six main phyla of which the coniferophytes (conifers) and angiospermophytes (flowering plants) are the most advanced and the most dominant.

> Green plants are capable of movement, reproduction, sensitivity, growth, respiration, excretion and nutrition though many of these characteristics and life processes are not at all obvious.

> Green plants are autotrophs – they make their own food by photosynthesis.

> Plant cells are usually characterised by having cellulose cell walls, chloroplasts (green tissue cells) and large, permanent vacuoles.

> Flowering plants have well-developed roots, stems, leaves and flowers.

> Sexual reproduction in flowering plants takes place in five stages including pollination, fertilisation, seed formation, seed dispersal and germination.

> Plant health is most easily affected by the availability of water, light and minerals, invertebrate organisms and plant pathogens including fungi, bacteria and viruses.

M-LEVEL EXTENSION > > > > M-LEVEL EXTENSION > > > >

Collect some samples of science work from children in Key Stage 1 or 2. Track through their work on the functioning of green plants. Compare this aspect with samples of work from children in other year groups to see the progression. If you can gain access to the work of one particular child over time, track the development of their understanding of the functioning of organisms.

REFERENCES REFERENCES **REFERENCES** REFERENCES REFERENCES

Bianchi, L. (2000) So what do you think a plant is? *Primary Science Review*, 61, 15–17.

Briten, E. (2006) Sowing the seeds of creativity. *Primary Science Review*, 91, 22–25.

Fernandes, F. M., Carvalho, M. and Silveira, M. (2006) Real trees in the classroom. *Primary Science Review*, 94, 9–11.

Heath, L. (2008) Teaching Life Processes. *Primary Science Review*, 101, 36–37.

Jewell, N. (2002a) What's inside a seed? *Primary Science Review*, 75, 12–15.

Jewell, N. (2002b) Examining children's models of seeds. *Journal of Biological Education*, 36(3), 116–23.

FURTHER READING FURTHER READING **FURTHER READING** FURTHER READING

DfE (2011) *Teachers' Standards*. Available at www.education.gov.uk/publications.

Dorling Kindersley Multimedia. *The Encyclopedia of Nature*. An excellent, information-packed, interactive CD-ROM with video clips and spoken text. Also available as an eBook.

Hollins, M. and Whitby, V. (2001) *Progression in Primary Science: a Guide to the Nature and Practice of Science in Key Stages 1 and 2*. London: David Fulton. Provides useful information on teaching strategies and, as the title suggests, children's progression in all areas of National Curriculum science.

Sharp, J. (ed.) (2004) *Developing Primary Science*. Exeter: Learning Matters. Provides useful information on all aspects of science education.

3
Functioning of organisms: humans and other animals

Curriculum context
National Curriculum programmes of study

At Key Stage 1, children should be taught about the differences between things that are living and things that have never been alive and to relate simple life processes to themselves and to other animals found in the local environment. They should also be taught about the main external parts of the human body, how their bodies work in general terms, and what it means to be healthy. Similarities and differences between children, and between animals, should be used to introduce variation and classification.

At Key Stage 2, children should be taught about specific life processes common to humans and other animals and to make links between life processes in familiar animals and the environments in which they are found. They should also be taught about some of the main internal organs of the human body, how their bodies work in detail, and how to stay healthy. Keys should be introduced as a means of classifying and identifying locally occurring animals.

Early Years Foundation Stage

Children must be supported in developing the knowledge, skills and understanding that will help them to make sense of the world. Their learning must be supported through offering opportunities for them to encounter creatures in their natural environments and in real-life situations and undertake practical 'experiments'. By the end of the EYFS, children should:

- investigate objects and materials by using all of their senses as appropriate;
- find out about, and identify, some features of living things they observe;
- look closely at similarities, differences, patterns and change;
- ask questions about why things happen and how things work.

Introduction

Children of all ages grow up fascinated and curious about their own bodies and the bodies of people around them. Children are also curious about other animals, particularly in terms of what they look like, where they live and how they behave. Since ancient times, philosophers and scientists have worked tirelessly to unravel the mysteries of how humans and other animals 'work' and to find ways of sorting and classifying them, with varying degrees of success. Today, despite living in a complex biological world seemingly dominated by advances in DNA studies and genetic engineering, such matters are relatively well understood.

Humans and the animal kingdom

Most of the living organisms known today have been systematically classified and placed into one of five major groups or taxa known as kingdoms. With the exception of viruses, the classification of which remains problematic, these include the following.

For more information on humans in general, see Chapter 4 on continuity and change.

- Animalia.
- Plantae.
- Protoctista.
- Monera.
- Fungi.

Humans belong to the animal kingdom (see Figure 3.1). The animal kingdom contains a particularly diverse group of complex, multicellular, eukaryotic organisms, all of which have a nervous system of one form or another, most of which are capable of locomotion at some stage in their life cycle, and most of which reproduce sexually. Animals are heterotrophs and get their food by eating plants and other animals. At other taxonomic levels, humans are **chordates** (phylum), mammals (class), primates (order) and hominids (family). Biologically, humans are also referred to as *Homo sapiens* (genus and species). Binomial terms such as this are assigned to all living organisms by international agreement and convention.

RESEARCH SUMMARY RESEARCH SUMMARY RESEARCH SUMMARY **RESEARCH SUMMARY**

In two early and classic studies by Bell (1981) and Bell and Barker (1982), only about 25 per cent of the 7-year-olds and 60 per cent of the 11-year-olds involved considered humans to be animals. This situation persists even today. At home and in school, animals are usually discussed within the contexts of domestic pets, farms or zoos. As humans do not have four legs, fur or generally make animal noises, they are not considered animals. Errors and misconceptions can be transmitted in the most unintentional of ways. The concept of 'animal' is clearly one that requires some attention and careful development. This concern persists in secondary education, as described by Day and Simms (2007).

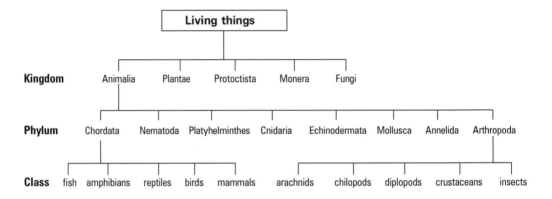

Figure 3.1 Partial classification of humans and other animals

Life processes

In common with all other living organisms, humans display characteristics which demonstrate that they are alive and carry out certain processes in order to stay alive. These include:

- movement;
- respiration;
- sensitivity;
- growth;
- reproduction;
- excretion;
- nutrition.

Organisms which are alive are said to be animate. Organisms which were once alive but are now dead, and things which have never been alive at all, are said to be inanimate.

RESEARCH SUMMARY RESEARCH SUMMARY RESEARCH SUMMARY **RESEARCH SUMMARY**

Piaget (1929) found that many young children considered bicycles, clocks, the Sun, the Moon, clouds and fire to be alive (animism), sometimes going as far as to assign human qualities or characteristics to them (anthropomorphism). It was noted by Piaget that life was readily, but often mistakenly, attributed to anything which simply appeared to move (by whatever means) or to make a noise or both. Later studies have demonstrated that even when correct distinctions are made between living and non-living things, children do not always use or apply the same life process criteria as biologists. Shepardson (2002) found that while they might refer to some of the same criteria, more complex factors such as dormancy, embryology, metamorphosis or dependency on energy transformations are rarely considered.

Movement

Movement occurs in humans in different ways and for different reasons. The most obvious movements take place during locomotion when they walk or run from one place to another. Walking and running are brought about by the action of muscles on bones in response to nerve impulses co-ordinated by the brain. Actions like walking and running are examples of voluntary movements – humans can choose to do them or not. Less obvious muscle movements take place as the heart beats, the chest expands drawing air into the lungs, and as food passes through the digestive system. Actions like these are examples of involuntary movements – they happen automatically without having to be thought about at all.

Respiration

Humans need energy in order to move, to grow, to carry out the reactions or metabolic processes which take place within cells, to transport substances around the body to where they are needed, and to maintain a constant body temperature of 37°C. This energy is made available in all living cells during respiration. Cellular respiration essentially involves a reaction between glucose, obtained from the digestion of food, and oxygen, obtained from air drawn into the lungs by breathing. Both glucose and oxygen are transported around the body to respiring cells by the blood. The waste products of **respiration**, carbon dioxide and respiratory water, are removed from respiring cells and breathed out. Cellular respiration is simply represented as a word equation:

For more information on respiration, see the section on converting energy in our bodies in Chapter 9.

$$\text{glucose} + \text{oxygen} \longrightarrow \text{carbon dioxide} + \text{water} + \text{energy}$$

Sensitivity

Humans are particularly sensitive to changes in the environment around them and to what goes on within their own bodies. The detection, interpretation and co-ordination of virtually all sensory information, together with any subsequent action or response, voluntary or involuntary, is made possible by nerves and the nervous system. Familiar human senses include sight, sound, balance, smell, taste and touch. These are made possible by specialised cells in the eyes (sensitive to light), the ears (sensitive to sound, gravity and motion), the nose and tongue (sensitive to chemicals in the air and in food) and the skin (sensitive to temperature, pressure and pain). Humans are also sensitive to the release of hormones into the blood from endocrine glands. The release of adrenaline from the adrenal glands located above the kidneys, for example, enables humans to react quickly to dangerous situations by speeding up reflexes and by increasing the rate at which the heart beats.

For more information on eyes and ears, see the section on how the eye receives light in Chapter 11, and the section on receiving vibrations at the ear drum in Chapter 12.

Growth

Humans grow and mature in order to complete their life cycle. Growth involves intellectual, physical, emotional and sexual development. Stages of growth include babyhood (about 0–2 years), childhood (about 2–11 years), adolescence (about 11–18 years) and adulthood (about 18+ years). As humans grow they do not simply expand and get bigger, their body cells divide and increase in number. Growth as a result of cell division takes place all over the body but at different rates and at different times. The energy required for growth and the substances required to make new cellular material come from food. Healthy growth therefore requires a healthy and balanced diet. As humans become old their cells are generally repaired and replaced rather than produced and they stop growing completely.

Reproduction

Humans reproduce and have children for all kinds of different reasons. Reproduction ensures that humans are replaced so that when they grow old and die the human species does not become extinct. Humans become capable of reproduction at puberty when the release of certain hormones brings about changes in secondary sex characteristics in males (about 12–15 years of age) and females (about 11–13 years of age). Humans reproduce sexually. The internal fertilisation of a female gamete or sex cell (an egg or ovum) by a single male gamete or sex cell (a sperm), usually following sexual intercourse, results in the fusion of cell nuclei and the formation of a single cell or zygote containing all the genetic material or DNA needed in order to divide and grow into a fully formed adult. Gestation in humans lasts about 40 weeks after which time a new baby is born.

Excretion

Excretion involves the elimination of unwanted waste products and potentially harmful substances associated with cells as they go about performing the metabolic processes that keep humans alive. Strictly speaking, excretion does not include the removal or egestion of undigested food as faeces since this material has never entered the bloodstream to perform any useful task within the body. Excretion mainly involves the kidneys, which filter out impurities and toxins within the blood, and the bladder, into which the kidneys drain. Most excreted waste leaves the body as urine (water and urea). Excretion also involves the lungs (where carbon dioxide is exchanged for oxygen) and the skin (sweat loss).

Nutrition

Humans need food for energy, to maintain their life processes and to remain in good health. The amount of food humans need depends on their age, body size, the work they do, lifestyle and even climate. The useful constituents within food are called nutrients. A balanced diet of carbohydrates, fats, proteins, vitamins, minerals, fibre and water provides all the nutrients humans need. During digestion, which takes place in the alimentary canal or gut running from the mouth to the anus, larger, more complex, insoluble food substances are broken down into smaller, simpler and more soluble food substances capable of passing through the intestinal wall of the gut into the blood. From there, they are eventually transported to where they are needed. The carbohydrate starch obtained from potatoes, bread, rice and pasta, for example, is digested to produce glucose which is used in cellular respiration.

Cells, tissues and organs: levels of organisation

Humans and other animals are made from cells. Cells are the building blocks and most basic units of life. Unlike most single-celled organisms which live independently and carry out all life processes on their own, the cells of complex, multicellular organisms like humans are highly specialised and carry out particular functions within the body (e.g. bone cells, muscle cells, nerve cells and red blood cells). This division of labour is reflected in different levels of organisation.

- A group of specialised cells of the same type that work together to carry out a particular function is called a tissue (e.g. bone, muscle, nerve).
- A group of tissues that work together to carry out a particular function is called an organ (e.g. heart, lung, stomach, brain, eye).

Despite specialisation, most cells within the human body have several features in common (see Figure 3.2). These include:

- a cell membrane;
- cytoplasm;
- a single nucleus and other organelles or 'mini–organs', including mitochondria and ribosomes.

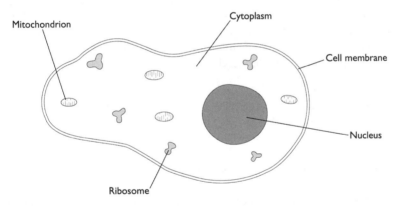

Figure 3.2 Human cell – common features (actual size 0.01 mm across)

The cell membrane protects and encloses the cytoplasm and nucleus and holds the cell together. Cell membranes are selectively permeable, allowing only certain substances to enter and leave. Cytoplasm is a jelly-like substance mostly made from water. Cytoplasm contains all of the cell's organelles or 'mini-organs', including the nucleus, and maintains the cell's shape. Human cells are eukaryotic. The nucleus contains a cell's genetic material or DNA. A cell's DNA determines what type of cell it is and controls what it does. The genetic material or DNA also has the ability to produce identical copies of itself in a process known as replication. This is important during cell division as humans grow. Other important organelles or 'mini-organs' include mitochondria, where glucose and oxygen react during respiration, and ribosomes, where amino acids are used to make proteins. Some cells may also

For more information on cells in general, see Chapter 4 on continuity and change.

contain granules of glycogen, a substance available as an energy store for use in times of need.

Major systems within the human body

The division of labour and levels of organisation evident in cells, tissues and organs extends to include the major body systems which also work together to carry out particular functions and the life processes described earlier.

RESEARCH SUMMARY RESEARCH SUMMARY RESEARCH SUMMARY **RESEARCH SUMMARY**

Children's ideas about the major systems that work together within their own bodies have been explored by Reiss and Tunnicliffe (2001) and Tunnicliffe (2004). While many of the older children involved could easily list a number of internal organs and other body parts, they often remained uncertain about their function and were usually unable to represent their size, shape and location accurately in drawings. The most obvious bones and muscles often appeared disconnected, the brain was often shown in isolation without any connecting nerves, and circulation was often considered to take place within a body filled with blood or within veins but not arteries. The digestive system was regularly portrayed as a simple tube within which food simply disappeared or passed through largely unchanged or in pieces.

The skeletal system

Humans possess an internal skeleton made largely of bone (see Figure 3.3). In mature adults there are about 206 bones in total. Babies may be born with up to 300, many of which fuse at a later time. Cuthbert (2000) confirms that children represent skeletons as individual and unrelated bones, indicating that the function of the skeleton is not well understood. The skeleton serves several purposes.

- It protects vital organs.
- It provides a framework which supports the weight of individuals and, with the help of muscles, allows humans to stand upright.
- It provides attachment for muscles and tendons, allowing free movements to take place across joints.
- It produces red and some white blood cells in bone marrow.

Despite appearances, bone is actually a living tissue composed of specialised bone cells or osteocytes existing within a matrix of hard, dry and relatively rigid calcium-rich deposits and a fibrous protein called collagen. Bone is penetrated by blood vessels and nerves which keep the bone cells alive. Bone cells are responsible for bone development and growth, and for bone repair and remodelling following accidents. They do this by absorbing calcium from the blood and from existing bone and redistributing it as required. About 99 per cent of the body's calcium is in bone. Most bones have hollow cavities filled with marrow. Bone marrow supplies the billions of red cells needed every day to replace those that are old and worn out. Bones are often described according to their shape and size.

- Flat bones (e.g. scapula, bones of the skull).
- Irregular bones (e.g. vertebrae).
- Short bones (e.g. bones of the wrists and ankles).
- Long bones (e.g. femur, humerus, ribs).

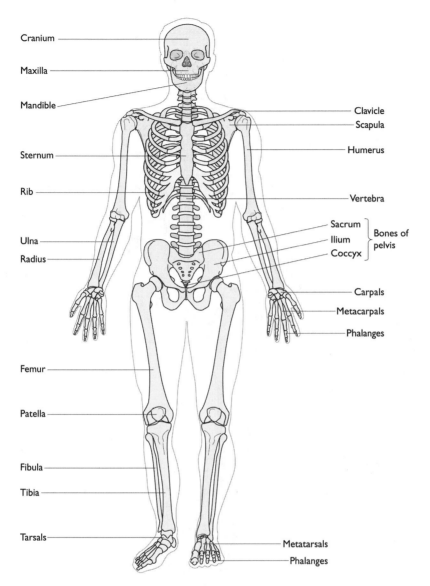

Cranium
Maxilla
Mandible
Sternum
Rib
Ulna
Radius
Femur
Patella
Fibula
Tibia
Tarsals
Clavicle
Scapula
Humerus
Vertebra
Sacrum
Ilium
Coccyx
Bones of pelvis
Carpals
Metacarpals
Phalanges
Metatarsals
Phalanges

Figure 3.3 Human skeleton

The human skeleton is jointed. Joints occur where two bones meet. These include:

- fixed or immovable joints (e.g. sacrum, pelvis and skull);
- partially movable joints (e.g. vertebrae, ribs and sternum);
- freely movable joints, including gliding joints (e.g. hands and feet), hinge joints (e.g. elbows) and ball and socket joints (e.g. shoulders).

Joints displaying movement are held together by ligaments. Wear at the surfaces of movable joints is prevented by layers of smooth cartilage or gristle (bone without the mineral salts). Some of the larger, freely movable joints are also lubricated by a viscous, synovial fluid.

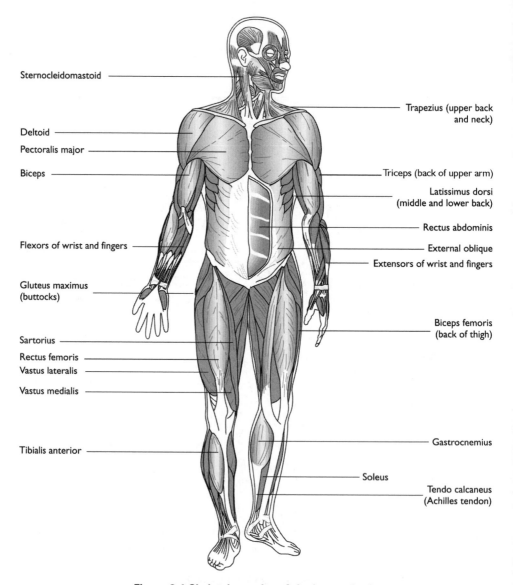

Figure 3.4 Skeletal muscles of the human body

The muscular system

Muscles have the ability to contract when stimulated by nerves. Almost all movements within the human body, voluntary and involuntary, are caused by muscles, and muscles, among other things, allow humans to locomote or to get around from one place to another. Muscles are grouped on the basis of structure and function as follows:

- skeletal muscle;
- smooth muscle;
- cardiac muscle.

The skeletal muscles (see Figure 3.4), lying immediately beneath the skin, are the most obvious and the most powerful muscles of the body and are responsible for most of the body's voluntary movements (those brought about by conscious thought). Skeletal muscle is made up of bundles of long muscle fibres which themselves are made up of elongated muscle cells. Skeletal muscles are attached to bones either directly or through tendons (a strong, inelastic, fibrous tissue). When skeletal muscles contract they pull on bones making them move at joints. Because skeletal muscles can only contract and pull, they work in pairs against each other. These are known as antagonistic pairs. As the biceps muscle contracts, for example, the arm bends or flexes at the elbow (the antagonist, the triceps muscle, is said to relax). When the triceps muscle contracts, the arm straightens or extends at the elbow (the antagonist, the biceps muscle, is said to relax). Smooth muscle is made up of layers of elongated muscle cells rather than bundles of long muscle fibres. Smooth muscle is responsible for most of the body's automatic or involuntary movements such as those associated with the alimentary canal. Contractions of smooth muscle in the alimentary canal push food through the digestive system with a wave-like motion known as peristalsis. Cardiac muscle is unique to the heart. Cardiac muscle is made up of branching fibres rather than bundles of long fibres or layers. Automatic or involuntary contractions of cardiac muscle pump blood around the body.

All muscles require energy to work. This energy is made available within individual muscle cells during respiration. Glucose and oxygen are delivered to the muscles by a network of blood vessels. If the demand for energy is too great, muscle cells may respire in the absence of sufficient oxygen resulting in the build-up and accumulation of lactic acid, which can lead to heavy and achy limbs and even cramp.

The nervous system

The nervous system works to control and co-ordinate almost every action that the body performs. It does this in response to both external and internal stimuli. The entire nervous system consists of specialised nerve cells called neurones. Neurones have the ability to transmit nerve impulses very rapidly (about 100 metres per second). The nervous system is divided into two main parts.

1. The central nervous system (CNS).
2. The peripheral nervous system (PNS).

The central nervous system includes the brain and spinal cord. It receives sensory information from the peripheral nervous system, processes it, and makes decisions which are then transmitted back through the peripheral nervous system as instructions for action. Humans have a complex and highly developed brain with different sensory areas responsible for particular processing functions. In addition to co-ordinating almost all of the body's actions, the brain is the site of all higher mental functions, including language, cognition and learning, reasoning, emotion and personality.

The peripheral nervous system consists of bundles of nerve fibres or nerves which lead to and from the central nervous system. Twelve pairs of cranial nerves operate the head and neck and emerge directly from the brain itself, while 31 pairs of spinal nerves from the spinal cord connect to the rest of the body. Within the peripheral nervous system there are two types of neurone.

1. Sensory neurones.
2. Motor neurones.

For more information on how the eye receives light, see Chapter 11.

Sensory neurones carry information as nerve impulses from receptors in sensory organs to the central nervous system. Receptors are groups of cells that are sensitive to a particular stimulus. In the eye, for example, the receptors are the rods and cones in the retina which are sensitive to light. Motor neurones carry instructions as nerve impulses from the central nervous system to effectors. Effectors are usually muscles or glands which respond to stimuli according to the brain's instructions and bring about a voluntary or involuntary response. The part of the peripheral nervous system that deals mostly with external stimuli and voluntary response (e.g. choosing to word-process an essay or to turn the pages of a book) is often referred to as the somatic nervous system. The part of the peripheral nervous system which deals mostly with internal stimuli and involuntary response (e.g. moving food along the alimentary canal) is often referred to as the autonomic nervous system.

A summary of how the nervous system works is presented as follows:

$$\text{stimulus} \Rightarrow \text{receptor} \Rightarrow \text{sensory neurone} \Rightarrow \text{CNS}$$

$$\text{CNS} \Rightarrow \text{motor neurone} \Rightarrow \text{effector} \Rightarrow \text{response}$$

Specific types of involuntary response called reflex actions are designed to protect humans from harm. In reflex actions, particular stimuli send information to the spinal cord from where it travels back to an appropriate effector in a reflex arc with little or even no brain involvement.

The endocrine system

Like the nervous system, the endocrine system works to control and co-ordinate certain actions that the body performs and regulates some of its many functions and processes. In response either to instructions from the brain or to changes in body chemistry itself, glands within the endocrine system release substances called hormones into the blood. These hormones or 'chemical messengers' target specific cells within specific organs to bring about change. The nervous and endocrine systems are linked via the pituitary or master gland located at the base of the brain. The pituitary gland releases a wide range of hormones which control the activity of other glands. Some effects of important hormones are described as follows:

- adrenaline from the adrenal glands above the kidneys speeds up the heart and breathing rates in response to fear or anxiety;
- oestrogen and progesterone from the ovaries initiate the onset of secondary sex characteristics in females at puberty and regulate menstruation and ovulation;
- testosterone from the testes (testicles) initiates the onset of secondary sex characteristics in males at puberty and regulates sperm production;
- insulin from the pancreas in the upper abdomen regulates blood sugar levels;
- thyroxine from the thyroid glands in the neck regulates body metabolism and overall growth.

Other hormones and hormone-like substances are responsible for a wide range of features including body rhythms (e.g. sleep), pain control and sexual attraction.

The respiratory system

The respiratory system plays its part in cellular respiration by ensuring that a regular supply of oxygen is made available to respiring cells and that the waste products (carbon dioxide with some respiratory water) are removed. The respiratory system involves:

- breathing or ventilation;
- gas exchange;
- cellular respiration itself.

The oxygen required for respiration comes from the air humans breathe. Breathing or ventilation proceeds by inhalation and exhalation. Inhalation involves drawing air into the lungs. Exhalation involves forcing air out of the lungs. The lungs themselves contain no muscle tissue. They are, however, attached to the ribs and diaphragm by pleural membranes. In order to draw air into the lungs, the diaphragm, a wall of muscle separating the thorax from the abdomen, and the intercostal muscles of the ribs contract causing the rib cage to expand. During exhalation, the diaphragm and the intercostal muscles simply relax. Adult humans breathe in and out about 15 times a minute. This respiration rate increases during periods of physical activity when the body's demand for oxygen is at its greatest and decreases during periods of rest and sleep.

Once breathed in, air reaches the lungs first through the mouth and nose and then via the trachea or airway which divides into two branches or bronchi. Each bronchus divides further into smaller bronchioles which end in millions of alveoli or air sacs. The alveoli are surrounded by a network of fine blood vessels or capillaries. Within the alveoli, oxygen from inhaled air and carbon dioxide in the blood, together with a small amount of respiratory water, are exchanged. From the lungs, the freshly oxygenated blood returns to the heart from which it is pumped out around the body to where cellular respiration is taking place.

The circulatory system

The circulatory system (see Figure 3.5) transports oxygen, food substances and heat all around the body. It gets white blood cells and platelets to where they are needed for the fight against disease and for repairing wounds. The circulatory system also collects unwanted waste products and other harmful substances from around the body and transports them to where they can be processed or removed. The circulatory system consists of three main elements.

1. The heart.
2. Blood vessels.
3. Blood.

Blood is circulated around the body by the heart, a four-chambered organ that works like two pumps side by side. Each pump forces blood around one of two connected circuits:

- the heart–lungs–heart or pulmonary circuit;
- the heart–body–heart or systemic circuit.

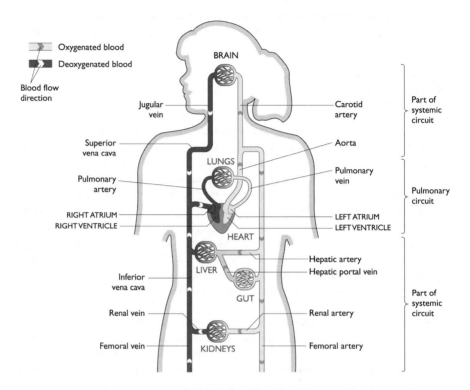

Figure 3.5 Human circulatory system (simplified)

In the pulmonary circuit, blood depleted in oxygen (deoxygenated blood) is pumped from the right ventricle of the heart through the pulmonary artery to the lungs where carbon dioxide is exchanged for oxygen. From the lungs, blood rich in oxygen (oxygenated blood) returns through the pulmonary vein to the left atrium of the heart. In the systemic circuit, oxygenated blood is pumped from the left ventricle of the heart through the aorta to the body. From the body, deoxygenated blood returns through the venae cavae to the right atrium of the heart. Blood is prevented from flowing backwards in the heart by valves. The rhythmic pumping of the heart is controlled by specialised pacemaker cells within the cardiac muscle which are sensitive to the expansion of the heart chambers as they fill with blood. When full, nerves stimulate the cardiac muscle to contract. The pumping of the heart is easily heard as a heart beat and the rush of blood through certain arteries can be felt as a pulse. An adult human heart beats about 60–80 times per minute. This heart rate increases during periods of physical activity when the body's demand for oxygen and glucose is greater than normal. The heart beats more slowly during periods of rest and sleep.

There are three types of blood vessel.

1. Arteries.
2. Veins.
3. Arterial and venous capillaries.

Arteries are strong, thick-walled vessels with narrow channels or lumens. Most arteries run deep within the body to protect them from damage. Arteries within the wrist and the neck do not (pulse points). Arteries conduct blood at high pressure away from the heart. Usually this blood is oxygenated. The pulmonary artery is a notable exception. Veins have thinner walls and wider channels than arteries and possess valves to prevent blood from flowing in the wrong direction. Veins conduct blood at low pressure towards the heart. Usually this blood is deoxygenated. The pulmonary vein is a notable exception. Away from the heart, arteries and veins form a more penetrating network of thin blood vessels known as arterial and venous capillaries.

Blood is a mixture of four main components.

1. Red blood cells or erythrocytes (produced in bone marrow).
2. White blood cells including phagocytes and lymphocytes (some produced in bone marrow, others in the body's lymphatic system and stored in the spleen).
3. Platelets or thrombocytes.
4. Plasma.

Red blood cells carry oxygen from the lungs around the body to where it is needed. The oxygen is attached to iron in the respiratory pigment haemoglobin. The red blood cells are responsible for blood's red colour. White blood cells provide natural immunity and defence against dangerous micro-organisms or pathogens and fight off disease by either 'eating' them (phagocytes) or by producing antibodies and antitoxins to counteract their presence (lymphocytes). Platelets are cell fragments which help blood to clot around and seal off wounds. Plasma is a straw-coloured liquid which contains all of the other blood components and just about everything the body needs to survive dissolved within it.

The digestive system
The digestive system (see Figure 3.7) essentially receives and, where necessary, processes food into substances which will dissolve. These are eventually transported around the body in the blood to wherever they are needed. In detail:

- carbohydrates like the starch in bread, potatoes, rice and pasta and the sucrose in sweet drinks are broken down to simple sugars like glucose which are used during cellular respiration;
- lipids or fats from butter, margarine, cooking oils, nuts, fatty meats and oily fish are broken down to glycerol and fatty acids which are used in cell membranes, for insulation and as a store of energy;
- proteins from milk, eggs, meat, fish and lentils are broken down to amino acids and used to manufacture new proteins which are used in building cells and making enzymes;

- vitamins from green vegetables, carrots, fresh fruit and liver are used in association with enzymes to control certain reactions such as cell regeneration, respiration, clotting, healing wounds and mineral absorption;
- minerals from table salt, milk, cheese, liver and sea food are used to ensure that nerves and muscles function well, that bones and teeth are hardened, that the haemoglobin in red blood cells is formed and that hormones are produced;
- water from drinks and other food sources is used in the cytoplasm of cells, blood plasma and maintaining the body's overall water balance so as to prevent dehydration;
- largely indigestible fibre from fresh fruit and vegetables and whole-grain products is used to provide roughage and bulk to stretch the wall of the intestine, to promote the passage of faeces and to keep the large intestine or bowel clean.

The processing of food takes place in three main stages.

1. Ingestion.
2. Digestion.
3. Egestion.

Ingestion occurs when food enters the mouth. Here it is chewed and moistened with saliva which contains enzymes. Teeth are important during ingestion (see Figure 3.6). Adult humans have 32 permanent teeth which replace an earlier set of 20 milk teeth. These include:

- incisors for biting and cutting;
- canines for piercing and tearing;
- premolars for grinding and crushing soft food;
- molars for grinding and crushing hard food.

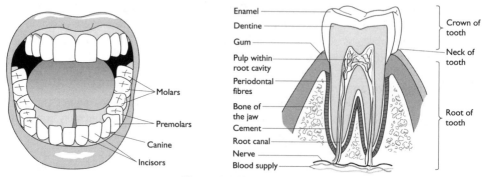

Figure 3.6 Human teeth

Digestion takes place in the digestive tract or alimentary canal, a 9-metre long tube from the mouth to the anus. The alimentary canal includes the oesophagus, stomach, small intestine (duodenum and ileum) and large intestine (caecum, colon, rectum and anus). Food is kept moving along the alimentary canal by the involuntary action of smooth muscles which line its walls. The contraction and relaxation of these muscles produces a wave-like motion referred to as peristalsis. Throughout most of the alimentary canal, food is mixed with a variety of digestive juices including hydrochloric acid, more enzymes and bile. Soluble products are then absorbed through the wall of the small intestine into the blood. The millions of projections or villi which line the inner wall of the small intestine, each supplied with its own network of arterial and venous capillaries, ensure that absorption is rapid. Most products of digestion are first taken to the liver for further processing. Largely undigested food continues onwards into the large intestine where the water content of the digestive juices released into the alimentary canal is reabsorbed by the body. This helps maintain the body's water balance and prevents severe dehydration. The remaining faeces are stored for periodic release or egestion from the anus.

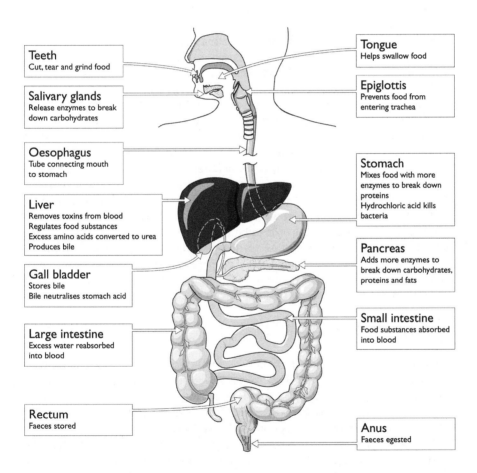

Teeth
Cut, tear and grind food

Salivary glands
Release enzymes to break down carbohydrates

Oesophagus
Tube connecting mouth to stomach

Liver
Removes toxins from blood
Regulates food substances
Excess amino acids converted to urea
Produces bile

Gall bladder
Stores bile
Bile neutralises stomach acid

Large intestine
Excess water reabsorbed into blood

Rectum
Faeces stored

Tongue
Helps swallow food

Epiglottis
Prevents food from entering trachea

Stomach
Mixes food with more enzymes to break down proteins
Hydrochloric acid kills bacteria

Pancreas
Adds more enzymes to break down carbohydrates, proteins and fats

Small intestine
Food substances absorbed into blood

Anus
Faeces egested

Figure 3.7 Human digestive system (simplified)

The reproductive system

While the biological purpose of reproduction in humans is to ensure the survival of the species, the emotional expression of affection between the individuals involved is as much a part of human sexual behaviour as the act of intercourse itself. Nevertheless, human reproduction can be summarised as follows:

- ovulation and the release once a month of a female gamete or sex cell (an egg or ovum) from an ovary into one of two Fallopian tubes;
- the sexual arousal of a male and a female;
- sexual intercourse, during which time the erect penis of the male is inserted into the vagina of the female and male gametes or sex cells (sperms) are released during an emission or ejaculation of semen;
- fertilisation, when ovulation and sexual intercourse coincide, resulting in the production of a zygote formed from the fusion of nuclei from the male and female gametes;
- the development of an embryo from the zygote as it divides and grows in the Fallopian tube on its way to the uterus;
- implantation of the embryo into the wall of the uterus about seven days after fertilisation;
- the development of a foetus from the embryo following the formation of major organs;
- birth after a gestation period of about 40 weeks.

The ovaries within the reproductive system of the female usually only release one single egg each month. Should fertilisation not take place, the egg and blood from the lining of the uterus are lost from the body during menstruation. In a single emission or ejaculation of semen during sexual intercourse, as many as 250 million sperm cells can be released. Only a few hundred may survive to reach the egg or ovum in the Fallopian tube and only one will penetrate its outer membrane to allow fertilisation to take place. Within the uterus, a human embryo and foetus grows within its own environment in a fluid-filled sac or amnion which protects it. A developing foetus cannot eat or breathe on its own. It gets its oxygen and other useful substances from its mother via the placenta and the umbilical cord. The placenta allows the blood systems of the mother and foetus to come close together without actually mixing. Hormones released during pregnancy bring about several changes in the mother in preparation for birth and birth itself.

REFLECTIVE TASK

What is an animal? This question is a lot trickier to answer than you might imagine, regardless of whether you are 5, 7 or 11 years of age. Are humans animals? What criteria would you use to demonstrate that they (we) are? What criteria might you expect children to use? Throughout the primary years, children of all ages are exposed to various aspects of the human body and the bodies of other animals and how they work. But for health and safety, as well as a whole host of other reasons, it is not always acceptable to bring body parts into class. How might you overcome such restrictions when teaching a class of 5-, 7- or 11-year-olds? Do you consider the use of secondary sources acceptable in terms of how you might define good primary practice? You will often hear primary teachers talk about the 'hands-on' nature of primary science. But what do you think they mean by 'hands-on' and is it the only or the most effective way of teaching? Is it too limited an approach? Whatever happened to 'minds-on'?

Health

The general health and state of well-being of humans can be adversely affected in many different ways, including:

- a poor diet;
- exposure to disease-causing micro-organisms or pathogens (germs);
- exposure to harmful substances;
- lack of exercise, rest and sleep;
- stress.

RESEARCH SUMMARY RESEARCH SUMMARY **RESEARCH SUMMARY RESEARCH SUMMARY**

Gill *et al*. (2006) found that primary children could hold quite complex notions about food and its consequences for health. However, while their knowledge about healthy and unhealthy food was generally good, it did not appear to influence their eating habits and food choices. Wyvill (2005) has provided some insight into how drugs education can be promoted with primary-aged children.

In many instances, the circumstances surrounding ill-health are beyond the control of the individual (e.g. war, famine, flood and drought).

Poor diet

A balanced diet is one which provides humans with all they need in order to avoid diet-related health disorders.

- Excess carbohydrate and fat can lead to obesity and coronary heart disease, while insufficient carbohydrate and fat can lead to overall body wastage and listlessness.
- Insufficient protein can lead to weak muscles, flabbiness and bloatedness.
- Mineral deficiencies can lead to problems associated with the nervous and muscular systems (sodium and potassium), the hardening of bones and teeth (calcium and phosphorus), acute tiredness and anaemia (iron) and overall growth (iodine for the hormone thyroxine).
- Vitamin deficiencies can lead to problems associated with night vision (retinol or vitamin A), nerve degeneration, blurred vision, cracked skin and stunted growth (thiamine or vitamin B1 and riboflavin or vitamin B2), poor wound healing and scurvy (ascorbic acid or vitamin C), soft bones and rickets (calciferol or vitamin D) and blood clotting time (phylloquinone or vitamin K).
- Insufficient fibre can lead to constipation and various bowel disorders, including cancer.
- Insufficient water can lead to dehydration and urinary infections.

Pathogens

Disease-causing micro-organisms or pathogens (germs) can be transmitted to humans in several ways including eating contaminated food, drinking contaminated water, coughing and sneezing, direct contact with another individual, sexual intercourse, and by vectors or disease-carrying organisms such as mosquitoes and fleas. Once beyond the body's external and internal defences or immune system, many pathogens damage or kill healthy cells as they feed and multiply or by releasing poisonous substances or toxins. Pathogens and the diseases they cause include:

- bacteria (e.g. sore throats, bacterial meningitis, tetanus, tuberculosis, cholera, typhoid, whooping cough and salmonella);
- viruses (e.g. colds, flu, viral meningitis, measles, mumps, polio, hepatitis, herpes and AIDS);
- fungi (e.g. athlete's foot, ringworm and thrush);
- protista (e.g. amoebic dysentery, sleeping sickness and malaria).

Many bacterial infections can be treated with antibiotics. Vaccination alerts the human body's internal defence mechanisms to certain types of bacteria and **viruses** in advance of exposure and can provide a degree of artificial immunity. A wide range of treatments are available to combat problems associated with **fungi** and **protista**. Strict adherence to basic rules of food preparation and storage, personal hygiene, and care when travelling to geographical areas known to have particular health risks may help to protect against and to avoid possible infection in the first instance. While not all diseases are treatable, many are preventable.

Harmful substances

Exposure to any of the harmful substances mentioned here, even over a short period of time, can lead to an increased tolerance of their effects, resulting in both chemical or psychological dependency and addiction. Such substances include:

- tobacco;
- alcohol;
- solvents;
- other drugs.

Smoking tobacco has been directly associated with breathing disorders (e.g. emphysema, bronchitis), blocked arteries, heart disease, lung cancer and nerve damage. Alcohol abuse and 'sniffing' solvents (the substances which give off fumes in glues, paints, polishes, hairsprays and cleaning products) have been directly associated with impaired performance, personality changes and damage to major organs including the lungs, liver, kidneys, stomach, heart and brain. The misuse of other drugs (e.g. depressants, stimulants, hallucinogens and painkillers) has also been directly associated with impaired performance, personality changes and damage to major organs. The transfer of pathogens (e.g. hepatitis, HIV) is possible if using or sharing dirty needles to administer drugs intravenously.

Lack of exercise, rest and sleep

Regular exercise can make humans stronger and more efficient. A lack of regular exercise can lead to a lack of tone in muscles, inflexible tendons and joints, and reduced lung capacity. In some instances, a lack of regular exercise can lead to clogged arteries and high blood pressure, both of which have been associated with heart disease. Humans also need time to rest and sleep so that the body can repair and 'recharge' itself. Humans differ in the amount of rest and sleep they need. Some babies may sleep for most of the day while some adults may need as little as 5 or 6 hours. Sleep deprivation can lead to hormone imbalances resulting in stress, heightened anxiety and reduced decision-making capability.

Stress

Stress can be caused by a wide range of physical, emotional and environmental factors (e.g. too hot, too cold, too humid, too noisy, too bumpy, too crowded, bereavement). At work, stress is most apparent when there is a mismatch between what is being expected of individuals in whatever situations or capacities they find themselves in and their ability to satisfy that expectation. Difficult to diagnose because of the factors involved, symptoms of work-related stress can include neck and back pain, faintness, hyperventilation, palpitations, crawling skin and even reduced sex drive. The effects of long-term stress can result in anxiety and depression, mood changes, mental exhaustion and fatigue, weight and hair loss, stomach ulcers and, in extreme cases, lost immunity against infection. Stress management and counselling can help.

PRACTICAL TASK PRACTICAL TASK **PRACTICAL TASK** PRACTICAL TASK **PRACTICAL TASK**

At Key Stage 2, together with friends and colleagues, produce a database containing information about eye colour, hair colour, height, hand span, shoe size, the ability to roll the tongue, handedness and pulse. Which of these are categoric, discontinuous and continuous variables? In terms of graphs and charts, how are they best displayed? Using your data set, how many different investigative questions can you raise? Which sorts of data would it be ethically improper to collect? At Key Stage 1, talk about physical characteristics and allow the children to assign themselves to groups where eye or hair colour is the variable.

Other animals

RESEARCH SUMMARY RESEARCH SUMMARY **RESEARCH SUMMARY RESEARCH SUMMARY**

Surprisingly little is known about the development of children's ideas about animals other than humans. The areas that have received some attention include classification, growth and animal skeletons. In the two early and classic studies by Bell (1981) and Bell and Barker (1982), only about 30 per cent of the 7- and 11-year-olds involved considered worms and spiders to be animals. Prokop *et al.* (2007) report a study in which the ability of children to classify vertebrates and invertebrates was examined. Even though girls did well in classification tasks, they still drew bones inside invertebrates. Shepardson (2002) provides insights into children's understanding and classification of insects.

Vertebrates

Vertebrates are animals with backbones. Vertebrates are classified into five main groups or classes, all of which are located within Phylum Chordata. These include:

- fish;
- amphibians;
- reptiles;
- birds;
- mammals (including humans).

Fish, amphibians and reptiles are ectothermic (older terms include 'poikilothermic' and 'cold-blooded'). Their body temperatures vary with the temperature of their surroundings. Birds and mammals are endothermic (older terms include 'homeo-thermic' and 'warm-blooded'). Their body temperatures remain constant regardless of the temperature of their surroundings. The following descriptions are broad generalisations.

Fish

Fish live in water and 'breathe' by absorbing dissolved oxygen through gills. They propel themselves with fins which also aid stability. Fish reproduce sexually. Fertilisation usually takes place outside the bodies of females after they lay their eggs. Most common fish have bony skeletons and bodies covered in slimy scales. Bony fish possess a swim bladder which keeps them buoyant and stops them from sinking when they stop moving. Bony fish also possess a lateral line canal, a sensory organ which detects changes in water movements around them. Fish with skeletons made from cartilage include sharks and rays.

Amphibians

Amphibians like frogs and toads all have a soft and moist skin that is prone to desiccation which means that they live in damp or wet places. Amphibians reproduce sexually. Most lay their eggs in water. Fertilisation usually takes place outside the bodies of females as they lay their eggs. Life cycles vary. Frogs and toads, for example, metamorphose as they grow. The larvae or juveniles (tadpoles) swim using a tail and breathe through external gills. Adult frogs and toads breathe through lungs, have four limbs, and move by hopping or jumping.

Reptiles

Reptiles like lizards, turtles and snakes are covered in dry scales or bony plates which resist water loss. They breathe through lungs and most live on land. Perhaps the best-known reptiles, dinosaurs, are now extinct. Reptiles reproduce sexually. Fertilisation usually takes place inside the bodies of females, who subsequently lay eggs with soft, leathery shells. Lizards move by walking or running using their limbs, while snakes move by strong muscular contractions.

Birds

Birds have a skin covered in feathers, wings, a beak and legs with scales. They breathe through lungs. With the exception of bats (mammals), birds are the only vertebrates considered capable of true and unassisted flight (though not all birds fly). In order to fly, bird bones are 'hollow' and often fused to provide lightness and strength. Birds reproduce sexually. Fertilisation takes place inside the bodies of females, who subsequently lay eggs with hard shells. The eggs are usually incubated in nests to keep them warm.

Mammals

Land-living mammals have four limbs and skin covered in hair. They breathe through lungs. Mammals reproduce sexually. Fertilisation takes place within the body of females and much early development takes place within a uterus. Mammals give birth to live, often helpless, young, which feed on the mother's own milk. Most mammals exhibit a high degree of parental care. Monotreme mammals, like the duck-billed platypus, lay eggs. The young of marsupial mammals, like the kangaroo, develop from a foetal state in a pouch. Some mammals live or spend much of their lives in water (e.g. whales, dolphins and seals).

Invertebrates

Invertebrates are animals without backbones. Strictly speaking, the term invertebrate has no real taxonomic meaning as it includes organisms from totally unrelated phyla. Its informal use is retained here. The invertebrate phyla include:

- Arthropoda;
- Annelida;
- Mollusca;
- Echinodermata;
- Cnidaria (coelenterates);
- Platyhelminthes;
- Nematoda.

Arthropods are the largest and most diverse group of invertebrates. The five major classes include:

1. Insects (e.g. butterflies, beetles and bees; body divided into three parts – a head, a thorax and an abdomen; three pairs of legs which emerge from the thorax; single pair of antennae; many adult forms have wings);
2. Crustaceans (e.g. woodlice; body divided into three parts – a head, a thorax and an abdomen; five or more pairs of legs; two pairs of antennae);
3. Diplopods (e.g. millipedes; segmented bodies, no distinct head; one pair of antennae; two pairs of legs per segment; herbivores);
4. Chilopods (e.g. centipedes; segmented bodies, well defined head bearing poison claws; one pair of antennae; one pair of legs per segment; carnivores);
5. Arachnids (e.g. spiders; body has two parts – a cephalothorax or prosoma and opisthoma; four pairs of legs; no antennae or wings).

Among the most common invertebrates encountered are woodlice (e.g. *Oniscus asellus*, an arthropod, Class Crustacea), earthworms (e.g. *Lumbricus terrestris*, an annelid, Class Oligochaeta) and snails (e.g. *Helix aspersa*, a mollusc, Class Gastropoda). Many children are likely to have come across these.

Woodlice

Woodlice live in damp leaf litter, around rotting wood and under loose logs where conditions are dark and damp and where they can eat decaying vegetation. Woodlice are most active at night. As with all arthropods, woodlice possess a rigid exoskeleton and jointed limbs which they use to get around. The body is divided into three parts – the head, the thorax and the abdomen. The head has one pair of compound eyes, two pairs of sensory antennae (one pair obvious) and a mouth. The thorax consists of seven segments which protect vital organs. Each segment bears one pair of legs. In pill woodlice, the thoracic segments allow complete enrolment. All abdominal segments are fused. On the underside of the abdomen it is possible to see cream-coloured 'gills' through which woodlice breathe. Small projections at the end of the abdomen include the telson and two uropods. These are more highly developed and modified in other crustaceans (e.g. shrimps and lobsters). In woodlice, the telson and uropods are used for conducting water towards the 'gills' which, like the rest of the organism, must be kept moist.

Woodlice reproduce sexually. Sperm is transferred from male to female. Eggs are carried around by the female in a water-filled brood pouch on the underside of the abdomen until they hatch. Woodlice grow by moulting which involves shedding the exoskeleton at regular intervals.

Earthworms

Earthworms spend most of their lives in soil, swallowing and digesting the dead and decaying organic matter within it (egested material forms worm casts). They surface occasionally, usually at night, or after long periods of rain which cut off their oxygen supply. Although they do not have eyes, earthworms are sensitive to light. Earthworms breathe through their soft, moist skin which is prone to desiccation. In mature adults, a swollen saddle or clitellum is situated towards the front of the organism, the front also being the most pointed end. While most major organs also lie towards the front, the muscular, nervous, circulatory and digestive systems run the whole length of the body. Earthworms have a hydrostatic skeleton (support and movement are made possible by fluid pressure within the organism). Earthworms crawl and burrow by contracting circular muscles within each of the front end segments and extending the body. Chaetae or bristles anchor the front while long-itudinal muscles running the length of the organism contract, pulling the hind end along. Earthworms are hermaphroditic and have both male and female reproduc-tive organs. Self-fertilisation is rare and sexual reproduction involving mutual sperm transfer is more common. Earthworm eggs are deposited in a cocoon formed around the clitellum. The cocoon is later left in the soil.

Snails

Snails are found around garden plants and wooded areas where conditions are dark and damp and where there is a ready supply of food. Snails are soft-bodied animals with a hard, coiled shell. This shell contains and protects major organs. Snails have a well developed head bearing two pairs of tentacles (one pair with eyes) and a soft elongated foot. The head and foot are moist and slimy. The slime helps movement, prevents desiccation and acts as a deterrent to predators. Movement occurs as a result of waves of muscular contractions which are easy to see. Snails breathe via a pulmonary aperture which allows gases in and out of the shell. Snails are hermaph-roditic and possess both male and female reproductive organs. Self-fertilisation is rare and sexual reproduction involving mutual sperm transfer is more common. Snails lay their eggs in soil. Young hatch with a 'proto-shell' which develops more fully as they grow. Snails are most active at night.

PRACTICAL TASK PRACTICAL TASK **PRACTICAL TASK** PRACTICAL TASK **PRACTICAL TASK**

At Key Stage 2, making and using 'local' branching keys to identify invertebrate organisms is a popular classroom activity. List the readily observable features and questions you would use to sort and classify a range of typical invertebrates including woodlice, earthworms, snails and spiders. If you are not sure what these and other invertebrates look like, go out and observe some. You should consider health and safety and the care of the organisms involved. Are there any observational features and questions which are better than others? Compare your key with the keys of friends and colleagues. Are they similar or different?

Ask Key Stage 1 children to divide sets of animals into two groups, e.g. those with legs and those without. Play 'odd one out', where you have a set of three animals.

A SUMMARY OF **KEY POINTS**

> Humans are animals.

> The animal kingdom can be conveniently divided into vertebrates (e.g. fish and reptiles) and invertebrates (e.g. arthropods, annelids and molluscs) though the term 'invertebrate' should be used with care.

> Humans and other animals are capable of movement, reproduction, sensitivity, growth, respiration, excretion and nutrition.

> Humans and other animals are heterotrophs – they get their food by eating plants and other animals.

> Human and other animal cells usually include a cell membrane, cytoplasm, a single nucleus and other organelles.

> The cells, tissues and organs of humans and other animals work together to carry out the particular functions that keep them alive.

> The health of humans can be affected in a whole manner of different ways, including poor diet, exposure to pathogens, exposure to harmful substances, lack of exercise, rest and sleep, and stress (most of these things affect other animals too).

M-LEVEL EXTENSION > > > > M-LEVEL EXTENSION > > > >

Think about how children in the Early Years could start to learn about the functioning of humans and other animals. Could this aspect build on any of the traditional topics or themes used in Nursery and Reception, such as Ourselves, People who help us, Farm animals or Minibeasts? Or, should this area be planned for discretely? Discuss your reflections with experienced Early Years practitioners and source recent research on how this can be addressed in EYFS.

REFERENCES REFERENCES **REFERENCES** REFERENCES REFERENCES

Bell, B. (1981) When is an animal not an animal? *Journal of Biological Education*, 15(3), 213–18.

Bell, B. and Barker, M. (1982) Towards a scientific concept of animal. *Journal of Biological Education*, 16(3), 197–200.

Cuthbert, A. J. (2000) Do children have a holistic view of their internal body map? *School Science Review*, 82(299), 25–32.

Day, A. and Simms, S. (2007) Snakes aren't animals, they're worms: technical taxonomies and their effect on examination performance. *School Science Review*, 88(325), 105–10.

Gill, P., Stewart, K., Treasure, E. and Chadwick, B. (2006) School Children's understanding about food and health in Cardiff, UK. Paper presented at a conference. http://iadr.confex.com/iadr/

Piaget, J. (1929) *The Child's Conception of the World*. London: Routledge and Kegan Paul.

Prokop, P., Prokop, M. and Tunnicliffe, S. D. (2007) Effects of keeping animals as pets on children's concepts of vertebrates and invertebrates. *International Journal of Science Education*, 30(4), 431–49.

Reiss, M. J. and Tunnicliffe, S. D. (2001) Students' understandings of human organs and organ systems. *Research in Science Education*, 31(3), 383–99.

Reiss, M. J. and Tunnicliffe, S. D. (2002) An international study of young people's drawings of what is inside themselves. *Journal of Biological Education*, 36(2), 58–64.

Shepardson, D. P. (2002) Bugs, butterflies and spiders: children's understanding about insects. *International Journal of Science Education*, 24(6), 627–43.

Tunnicliffe, S. D. (2004) Where does the drink go? *Primary Science Review*, 85, 8–10.

Wyvill, B. (2005) What drugs have you taken this week? *Primary Science Review*, 86, 12–15.

FURTHER READING FURTHER READING **FURTHER READING** FURTHER READING

DfE (2011) *Teachers' Standards*. Available at www.education.gov.uk/publications.

Dorling Kindersley Multimedia. *The Ultimate Human Body*. An excellent, information-packed, interactive CD-ROM with video clips and spoken text.

Hollins, M. and Whitby, V. (2001) *Progression in Primary Science: a Guide to the Nature and Practice of Science in Key Stages 1 and 2*. London: David Fulton. Provides useful information on teaching strategies and, as the title suggests, children's progression in all areas of National Curriculum science.

Sharp, J. (ed.) (2004) *Developing Primary Science*. Exeter: Learning Matters. Provides useful information on all aspects of science education.

4
Continuity and change

Curriculum context

National Curriculum programmes of study

At Key Stage 1, children should be taught that humans and other animals produce offspring, that these offspring grow into adults, and that seeds grow into flowering plants.

At Key Stage 2, children should be taught about the main stages of the human life cycle and about the life cycle of flowering plants.

Early Years Foundation Stage

Children must be supported in developing the knowledge, skills and understanding that help them to make sense of the world. By the end of the EYFS, children should:

- look closely at similarities, differences, patterns and change;
- ask questions about why things happen and how things work.

Introduction

What makes us human and how do we pass this on to the next generation? What is our relationship, as a species, to the other animals which inhabit the planet? These questions are fundamental to understanding who we are. Genetics provides clues to these and many other questions about living things. In 1996, the 12 million chemical links in the genetic material of a strain of yeast were catalogued, and by the year 2001 the 3000 million chemical connections in the DNA of humans were listed. The use of this information and the introduction of genetically engineered crops are likely to be two of the main environmental issues of the twenty-first century.

Species

A species is a group of animals, plants or micro-organisms which share a wide range of common characteristics and can breed together to produce fertile offspring. **Evolution**, the process which leads to species formation, is continuous, so the boundaries between many species are not sharp, with crosses (hybrids) between species being possible.

Similar species are grouped together into a genus. In the case of *Homo sapiens*, there are no other species within the genus Homo. However, in the case of cats, our domestic cat, *Felis catus*, will readily interbreed with the ancestral form *Felis libyca* (the wild cat of Egypt). Other members of the cat genus include *Felis pardalis* (ocelot) and *Felis serval* (serval).

> **REFLECTIVE TASK**
>
> Think about where species and varieties fit into your understanding of animals and plants. Why has the vast array of different dog types not led to completely new species?

The formation of new species results from physical isolation of separate populations. **Darwin** studied the finches which had been blown from the South American mainland hundreds of thousands of years ago to the Galapagos Islands where they became isolated from the main population of finches. In their new homes the birds began to adapt to new conditions. The small, isolated pioneer population began to exploit the different sources of food and eventually diverged into different species. This selective breeding emphasised distinguishing characteristics leading to the formation of about a dozen new species, e.g. the cactus finch and the ground finch. Each species is a variation on the theme of the original birds.

For more information on species, see sectons later in this chapter for how species are formed.

Variation within species

The variation between individuals within the same species is striking. Some people, for instance, are large, some are small, some skin colours are black, others are white. The variation between people is so strong that we could recognise many tens of individuals from all the rest of the world's population. However, even the range of differences within the human species is dwarfed by the range of varieties within some animal species, such as dogs and horses.

Many of the measurable characteristics in a population show a predictable pattern in their distribution. Most individuals fall within a fairly broad middle band with only a few individuals at the two extremes. In humans this is true for features such as height, weight, hand span and intelligence. The resulting graph has the shape of a bell (see Figure 4.1 below).

Many species produce thousands of young, yet only two are needed to replace the parents. Why such profligacy? What is it which singles out so few from so many? It is the competition between individuals of the same species which results in evolutionary change. Only those individuals most suited to the environmental conditions of the time will survive. For instance, an individual finch with a small pointed beak, suited to picking insects out of bark, is more likely to survive a period in which the normal seed food is not available. A plant seedling, better suited than others of its generation to grow in dry conditions by virtue of a slightly more waxy leaf, will survive drought to pass on its genes to the next generation.

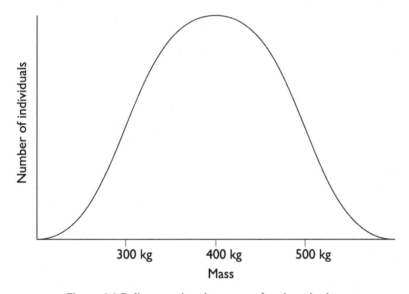

Figure 4.1 Bell curve showing mass of male polar bears

This selection of individuals can be illustrated by thinking of a population of bears living in the northern latitudes. Larger bears are more likely to survive cold conditions because they lose heat less rapidly than smaller bears. In this respect large size confers an advantage, so big bears survive to pass on their genes for largeness. The same group of bears have a variety of coat colour. In snowier conditions a white coat for camouflage is favoured. This leads to those bears with white coats surviving at the expense of those with brown coats. The change in size and coat colour, together with physical isolation of populations, can explain why the top predator of the northern land is the polar bear. So why didn't the size of polar bears simply carry on increasing leading to super-large polar bears? Size is restricted through competition within the bear population so that, in periods when there is severe competition for food, a huge bear would simply be outcompeted by its nimbler cousins. In addition, the sheer bulk of food that a super-large bear would need might not be available within one bear's territory.

PRACTICAL TASK PRACTICAL TASK **PRACTICAL TASK** PRACTICAL TASK

Talk with a breeder of dogs, cats or flowers. Discuss with them how they select for the characteristics they require.

DNA

DNA is the chemical which controls all the functions of the cell and thereby the complete organism. It is present in the simplest living things, such as bacteria, and the most complex, such as mammals. DNA is a long molecule made of two linked spirals. The links between the spirals are made by four chemicals which can be arranged in many different combinations to provide a code telling the organism how to develop. Each organism is made up of a unique sequence of these four chemicals arranged along the length of the DNA molecule. DNA is the basis of life because it can replicate itself, providing the same message to each cell in a body and to succeeding generations.

The human genome project set out to map the sequence of these four chemicals along the DNA molecule of a typical human. When it is completely understood, these myriad combinations of four chemicals will reveal the code for building a human being. In theory, it is possible to build an organism from basic chemicals once the sequence of the DNA has been understood. At the time of writing, understanding the implications of the lists of chemicals lags far behind the simple cataloguing. Announcements that the sequence for a particular characteristic has been identified are often followed by the realisation that to do anything meaningful with the information will take many years of research.

Genes

In classic genetics the **gene** is the basic unit of inheritance. In **Mendel**'s early work, genes were thought of as the units of heredity. In peas, the wrinkliness or colour of the seed are characteristics which are controlled by a gene. In humans, skin, eye and hair colour are controlled by a gene. However, in modern **genetic** theory a gene is regarded as a section of the DNA molecule which controls the production of a particular protein. It may take the interaction of many sections of DNA (several genes) to influence a particular characteristic of an organism. More complex characteristics, such as intelligence or the predisposition to violence in humans, are the result of many genes and their interaction with the environment.

The degree to which our genes determine the way we look and behave is still the subject of fierce debate. There is little doubt that the average height of people in rich countries is increasing. Better nutrition is contributing to people achieving close to their genetic maximum heights. These heights are still very varied. Educationalists argue that if we give children a richer education they are more likely to achieve to the limit of their intelligence imposed by their genetic inheritance.

Chromosomes

Chromosomes contain all the genetic information which allows cells to divide and produce identical replicas of themselves. A chromosome is actually one long DNA molecule and proteins. Chromosomes can be seen in the nucleus of each living thing as thread-like structures. In nearly every cell of the body they occur in pairs. One of each pair comes from the individual's mother and the other from their father. When there is a pair of chromosomes which look alike (though they are far from identical in detail since one is from each parent), the pair is referred to as being homologous. Each living thing has a characteristic number of chromosomes – in humans there are 23 pairs of homologous chromosomes, in peas there are seven and mice have 20 pairs. You will frequently see pictures of 46 long thin crosses looking like four-legged starfish. This is a microscopic view of the chromosomes just before they divide. Note that each chromosome in the pair is replicating itself separately and that there are actually 92 chromosome strands (chromatids) in the picture. The chromosome which has duplicated itself will unpeel from the duplicate and one will go into one daughter cell and the replica will go to the other.

The process of DNA replication

Have you ever wondered how the single cell of a human embryo knows how to make some cells become bone, others to become brain and still others to become skin? The body accomplishes this remarkable feat of differentiation and growth using the code book contained in the chromosomes. Each cell of the body needs a full set of codes to know how to grow and develop.

Living things grow and replace dead cells through the process of cell division. To accomplish cell division each new daughter cell must have all the genetic information which was available to the parent cell. When chromosomes are replicated in cell division, each chromosome builds a copy of itself so that when it splits each daughter cell will be identical. This is done by the chromosome building a copy of itself from chemicals in the cell. (See Figure 4.2.)

The process of making new daughter cells from parent cells is controlled by the chromosomes. In essence, new cell production follows this pattern:

- each chromosome makes an identical copy of itself;
- the new pairs then split apart, with one of each chromosome going to each of the new daughter cells (see Figure 4.3).

Cell division leading to the production of sex cells (eggs and sperm in humans) follows a path which leads to the production of sperm and egg cells containing only 23 chromosomes each. This is in contrast to ordinary cells which contain homologous pairs of chromosomes adding up to 46 in total. When the sex cells fuse to form a fertilised egg, the chromosomes from the sperm complement those of the egg to give the full complement of chromosomes.

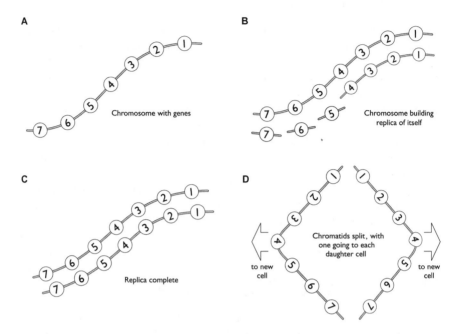

Figure 4.2 Chromosome replication

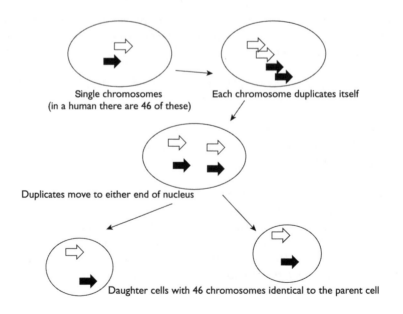

Figure 4.3 Cell division

Inheritance

Should children born to body builders be big and strong? Should the children born to criminals be devious and crooked? A heavily scarred person's child should be born scarred. We almost expect some of these strange ideas to be true – haven't they been borne out in reality? Did not J. S. Bach father a line of musically talented children? The idea that children **inherit** the characteristics that their parents developed during their lives is beguiling but entirely mistaken.

For more information on Mendel, see the section on genes above.

Another view which was commonly held before the work of Gregor Mendel, the father of genetics, was recognised at the end of the nineteenth century, suggested that when two lines of people or animals breed the resulting offspring would be a blend between the two. To explore this idea Mendel bred two pure lines of peas – one yellow and the other green. (It is possible to breed pure lines through a kind of plant incest.) He then interbred the two lines of peas. According to the prevailing idea of the time, the two characteristics of yellowness and greenness should have blended, resulting in peas that were greenish-yellow. Instead, Mendel was left with peas in a ratio of three yellow to one green. Mendel used the results to define the basic rule of inheritance. He proposed that the factor that led to yellow peas is **dominant** over the green factor.

These same rules apply to humans. The condition of albinism, where the individual lacks pigment in their body, occurs in one person in several thousand in England. In other parts of the world, in native Americans for instance, the prevalence is much higher at approximately one in one hundred and fifty. The gene for albinism is **recessive**, as is the gene for greenness in peas. Most people who carry the gene for albinism are completely unaware that they have this. Should a man carrying the recessive albino gene have a child with a woman also carrying the gene, the chance of their child showing albinism is one in four, just the same as the ratio of green peas to yellow.

Can you roll your tongue lengthways? If you can't do this then both the gene you get from your father and the one from your mother were for non-rolling. In other words, you have two copies of the same gene. For this gene you are **homozygous**. We can be sure that this is the case because the gene for rolling is dominant over the non-rolling gene. Individuals who either inherited the same gene for rolling from both parents (homozygous for rolling) or inherited one gene for rolling and one gene for non-rolling (**heterozygous**) will be able to roll their tongue.

Person A   rr (homozygous – is not able to roll)
Person B   RR (homozygous – is able to roll)
Person C   Rr (heterozygous – is able to roll)

Woman A (non-roller)			Man B (roller)		
r	r		R	R	
Rr	Rr		Rr	Rr	All children can roll tongue

Woman A (non-roller)		Man C (roller)	
r	r	R	r
Rr	Rr	rr	rr On average, 1 in 2 will not roll

REFLECTIVE TASK

You might track this characteristic in your own family ... but a word of warning ... you should think carefully before you do.

There are many diseases which are inherited through recessive genes. These include haemophilia, sickle cell disease and phenylketonuria. At present there is little that can be done to treat the underlying cause of these diseases, apart from helping would-be parents who are carriers of these diseases to make their own decision about whether they should have children. However, in many cases the diseases can be managed to give sufferers a good quality of life. For instance, phenylketonuria, which is an intolerance of a chemical contained in many foods, can be managed by careful diet. This allows sufferers to have an almost completely normal life.

So why bother with sex? It simply seems to lead to problems with **mutations** and inherited disease. The answer seems to lie in the need for large animals to keep one step ahead of bacteria and viruses. Since bacteria can reproduce every couple of hours, they can mutate and change very rapidly giving rise to new forms against which the majority of a population has no defence. AIDS, for example, has devastated whole populations, yet even in very high risk groups there are individuals who seem immune to the disease. It seems that there is something in their genetic make-up which offers them protection against the disease. If every single person in a population was exposed to HIV, then the only people to survive would be those with the abnormality which gave them protection. This chance resistance would, if the disease became a global pandemic, give these individuals a chance to survive and breed, thereby passing their resistance on to their children.

REFLECTIVE TASK

There are many inherited diseases which rule out people getting life assurance and medical insurance. Find out what you can about these exclusions and discuss the morality of such a policy.

Mutations and variation

For more information on replication, see the section on DNA above.

In the process of replication, mistakes are sometimes made by the body and DNA is copied incorrectly. This can be caused by chance, by chemicals in the environment or by radiation. Mistakes in copying genetic information are called mutations. These happen both in the process of cell division to form new body cells and in the production of egg and sperm. Errors in copying the chromosomes in new body cells can result in cancers, where cells multiply uncontrollably. Errors in making egg and sperm cells can lead to minor mutations or, in extreme cases, genetic diseases in offspring.

Faulty copying of DNA drives evolution. Most mutations in offspring are not notice-able or are completely benign. For instance, people with six fingers or toes are not uncommon and suffer no ill effects. Indeed, some mutations might even prove useful, leading to marginal advantages which prove decisive in ensuring survival when conditions change.

Genetic engineering

Normal selective breeding takes the genes from one variety of the same species or from two very closely related species and allows them to mix in unpredictable ways. Most of the crosses obtained in this way are useless to people and a few, such as one cross between varieties of potato, have proved to produce harmful effects on a food crop. Selective breeding, over thousands of years, has enabled farmers to pick the best features of one generation and pass them on to the next to produce better, tastier, heavier-yielding and faster-growing crops. Genetic engi-neering takes a radically different approach from normal selective breeding in the production of new varieties.

Scientists identify genes from one organism which have the code for a protein which is potentially useful to agriculture. These include the genes from a bacterium which make it resistant to weedkiller and the genes from a flat fish which make it resistant to freezing. These potentially useful genes are isolated and then inserted into the chromosome of the organism which will benefit from them. Once incor-porated into the genetic material of the new organism, the genes will be present in every cell. In the case of the weedkiller resistance, the new plant (Roundup ready soybeans, for instance) will tolerate weedkiller. In the case of the fish anti-freeze, the tomato which has the fish gene will not be damaged by freezing temperatures.

For more information on protein production, see the section on DNA above.

Clones

Clones are merely identical copies of a living thing. People have been making clones of plants for thousands of years. By merely pulling a side bulb from a daffodil, a tuber from a clump of potatoes or a side shoot from a spider plant, you are making a clone. Growing on the plant results in a new plant which is in all genetic respects identical to the parent. In sexual reproduction there is a completely novel individual made by combining the genes from two individuals. In the case of vegetative reproduction there is no combination. Clones of large animals such as mammals produced in the laboratories are recent innovations and are produced when one cell of an animal is persuaded to divide and then differentiate (form a variety of cell types). Identical twins are clones of each other since the egg from which both were formed divided after fertilisation leaving both twins with identical genes.

RESEARCH SUMMARY RESEARCH SUMMARY **RESEARCH SUMMARY RESEARCH SUMMARY**

Varville *et al.* (2005) surveyed the understanding that children aged 9–15 held about inheritance and technical vocabulary such as 'gene' and 'DNA'. Their results indicated that the majority of children could differentiate between biological inheritance and social factors that are not heritable. The majority had heard of DNA and genes but did not know much about them.

Jones (2005) was an NQT in a primary school who argued strongly that DNA and genetics should be taught explicitly in primary schools. She cited the pervasive discussions of DNA proof in courts and the central part played by genetics in such films as *Jurassic Park* and *Superman*. She suggests activities designed to interest primary school children in genetics.

Smith and Williams (2007) reviewed studies of how children's understanding of genes develops as they get older.

Evidence for evolution

For more information on Darwin, see the section on species at the start of this chapter.

When Darwin first put forward his theory of evolution through **natural selection**, the vast majority of people believed that God had created everything and that little, if anything, had ever changed. Today, most scientists believe that all living things are descended from ancestral forms and that the changes have happened over colossal periods of time.

The limbs of vertebrates, except fish, are based on the same basic plan. This indicates that all vertebrates are descended from a common ancestor, probably a fish. The limb bones of vertebrates can even be called by the same names: humerus, radius and ulna, and digits. Surprisingly, the flippers of whales and dolphins show the same basic vertebrate pattern.

Many animals and plants exhibit vestigial structures, which are parts of the body for which there is no apparent use today. The guts of humans and other meat-eating vertebrates still have an appendix, which in grass-eating animals is important in digesting cellulose. The presence of this vestigial structure indicates that all mammals are descended from a common ancestor. Similarly, humans still have the vestigial remains of a tail, the coccyx, which they share with many other mammals.

Fossil evidence for evolution is abundant, even though it represents a tiny fraction of the living things which have existed. The fossil record shows that before the Cambrian period, about 570 million years ago, the only animals were simple soft-bodied creatures such as sponges, worms and jellyfish. At the start of the Cambrian period, there is evidence of a widespread expansion of the number and variety of life forms. The fossil record is punctuated by mass extinction events where large numbers of animal and plant types became extinct in a short period. The causes of these extinctions remain contentious, with theories ranging from meteorite strike to huge volcanoes and alien viruses being suggested. The most famous mass extinction led to the demise of the dinosaurs about 65 million years ago. The loss of these huge and varied creatures opened up niches. There followed a very rapid evolution of birds and mammals to take the place of the huge reptiles.

- Ocean environments were filled by dolphins and whales in the place of plesiosaurs.
- The place of pterosaurs in the sky was taken by the birds.
- Vegetation feeders such as Diplodocus were replaced by large mammals such as cattle and horses.
- Predators such as Tyrannosaurus were eventually replaced by animals such as cats.

The evolution of some species can be followed in some detail as there are relatively abundant remains. Elephant and horse evolution can be traced through fossils. The first elephants, which appeared about 60 million years ago, were about the size of a pig with a small snout and tusks. Many types of elephant evolved from this including woolly mammoths, which were common in the ice age. Frozen mammoth remains are still to be found in permafrost. Cave paintings by people about 40 000 years ago show that they were found in Europe before becoming extinct about 15 000 years ago.

The geographic distribution of mammals gives further evidence for evolution. The evidence suggests that ancestral species were able to colonise different continents at different times and, once isolated, the ancestral species evolved as they adapted to the local conditions. The South American jaguar is strikingly similar in habits and build to the African leopard but they belong to distinct species. The South American llama is similar to the African and Asian camel.

For more information on evolutionary processes, see the section on mutations and variations above.

PRACTICAL TASK PRACTICAL TASK **PRACTICAL TASK** PRACTICAL TASK

Gather pictures of mammals such as gorillas, monkeys, dogs, cats and cows. Ask Key Stage 1 children which ones look most like people and which ones behave most like people. When working with Key Stage 2 children, debate the order of similarity by arranging pictures from those least like humans to those most like humans.

The evolution of humans

The evolution of humans can be traced from the common ancestor of all animals through fish and mammals (see Figure 4.4).

Humans are animals	We move fairly rapidly, and eat plants or other animals.
Humans are vertebrates	We have a backbone. A backbone is such a complex item that it is unlikely to have evolved more than once. This implies that all vertebrates are descended from a common ancestor.
Humans are mammals	We have hair, produce milk and have three separate bones in the ear. The three ear bones are believed to be the same three bones which make up the complex jaw of a reptile.
Humans are primates	We have fingernails not claws, an opposable thumb and four incisor teeth in the upper jaw and four in the lower jaw.
Humans are apes	We have a Y-shaped pattern on the molar teeth, shoulder blades at the back and we don't have a tail. (This is why monkeys are not apes.)

When trying to decide which are our closest relatives, we have to take care not to rely solely on appearance. In the case of the bat and a bird, for instance, the fact that they both fly is coincidental and is the result of parallel evolution. However, the other great apes, chimps and gorillas in particular, share many of our characteristics such as the shape and approximate size of skull and patterns on teeth. These animals also share a similar number of chromosomes (humans 46; chimps 48;

For more information on chromosomes, see the section on DNA replication above.

gorillas 48). There is considerable evidence to suggest that the chimp and human lines split about 8 million years ago.

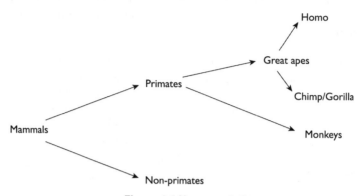

Figure 4.4 Human relatives

At any one time in the past there was more than one type of human-like animal. Even in the fairly recent past (about 35 000 years ago) there were at least two types of human being: modern humans and Neanderthal people. The genus name *Homo* is applied to animals which have many of the characteristics of modern people. The fossil evidence for humans is not extensive and many of the links are tentative. However, the overall trend is clear. Humans have evolved larger brains, an upright gait and increasing technical expertise.

For more information on cells, see the sections on cells, tissues and organs in Chapters 2 and 3.

Animal cells contain mitochondria which are responsible for the breakdown of sugars to release energy. The DNA within the mitochondria is passed on from mother to daughter in an unchanged line on the X chromosome (the female chromosome). This makes it possible to trace an unbroken path back through generations of females in the same way that European surnames allow you to trace descendants back through the male line. However, maternal inheritance has not led all women to have identical DNA. Through mutation, which happens at a regular and predictable rate, women now have different sequences within their mitochondrial DNA. Present-day African populations have the greatest variation in their DNA, leading to the conclusion that this population must have existed for a long time to allow the mutations to take place. Non-African populations show less variation because they are descended from small pioneer groups. This narrowness of the range of mitochondrial DNA is particularly marked in the native people of North America, whose ancestors were a very small group of Asians who crossed from Siberia.

So, are we going to continue to evolve as a species? The brutal view suggests that species evolve when there is a surplus of individuals struggling for limited resources. Those most suited to succeed in the struggle by virtue of a telling advantage will survive to pass on their genes to the next generation. Another view is that it will be ideas which evolve as we learn to control the physical environment.

A SUMMARY OF **KEY POINTS**

> A species is a group of organisms which can interbreed to produce fertile offspring.

> DNA controls the functioning of all cells.

> DNA is passed from parents to future generations.

> DNA is replicated in cells and mistakes in copying give rise to mutation.

> Genes are sections of DNA which determine characteristics of organisms.

> Chromosomes contain all the genetic information which allows cells to make replicas of themselves.

> Acquired characteristics are not inherited.

> Asexual reproduction leads to very little variation.

> Evolution occurs when changes in the environment select those organisms most suited to survival.

M-LEVEL EXTENSION > > > > M-LEVEL EXTENSION > > > >

Consider whether this complex aspect of science could be better understood by children if it was presented as part of a cross-curricular unit or topic based on continuity and change. Would it help to look at patterns in maths, similarities and differences in personal, social and health education and stories about parents and children as literacy texts? What about the historical background to DNA discovery or aspects of human geography?

REFERENCES REFERENCES **REFERENCES** REFERENCES REFERENCES

Jones, B. (2005) My Father's Ears. *Primary Science Review*, 86, 25–29.

Smith, L. A. and Williams, J. M. (2007) It's the X and Y thing: cross-sectional and longitudinal changes in children's understanding of genes. *Research in Science Education*, 37(4), 407–22.

Varville, G., Gribble, S. J. and Donovan, J. (2005) An exploration of young children's understanding of Genetic Concepts for Ontological and Epistemological Perspectives. *Science Education*, 89(4), 614–33.

FURTHER READING FURTHER READING **FURTHER READING** FURTHER READING

DfE (2011) *Teachers' Standards*. Available at www.education.gov.uk/publications.

Jones, S. (2000) *The Language of the Genes*. London: Flamingo. A highly readable account of the mechanisms of genetics, along with many fascinating facts about inheritance.

Jones, S. (2009) *Darwin's Island*. London: Little Brown Books.

Ridley, M. (2003) *Nature via Nurture*. London: Harper Collins. A highly readable account of the ways in which inheritance and environment interact.

5
Ecosystems

Curriculum context

National Curriculum programmes of study

At Key Stage 1, children should be taught how to find out about the different kinds of plants and animals in the local **environment**, and to identify similarities and differences between local environments and ways in which these affect animals and plants that are found there.

At Key Stage 2, children should be taught how locally occurring animals and plants can be identified and assigned to groups, that the variety of plants and animals makes it important to identify them and assign them to groups, about ways in which living things and the environment need protection, about the different plants and animals found in different habitats, how animals and plants in two different habitats are suited to their environment, to use food chains to show the feeding relationships in a habitat, about how nearly all food chains start with a green plant, and that micro-organisms are living organisms that are often too small to be seen, and that they may be beneficial or harmful.

Early Years Foundation Stage

With regard to their Knowledge and Understanding of the World, young children should: notice and comment on patterns; show an awareness of change; explain their own knowledge and understanding; and ask appropriate questions of others.

The Early Learning Goals suggest that young children should: investigate objects and materials by using all their senses as appropriate; find out about and identify some features of living things, objects and events they observe; look closely at similarities, differences, patterns and change; ask questions about how things happen and how things work.

In learning about places, young children should notice differences between features of the local environment; observe, find out about and identify features in the place they live and the natural world; find out about their environment; and learn about those features they like and dislike.

Introduction

Studying ecosystems is a way of helping children to make sense of interrelationships in the world around them. By considering feeding relationships through food chains and webs, we can help children to extend their ideas about food sources to beyond the local supermarket. By studying habitats and the interdependence of living things, we can help children to appreciate situations where life is not dependent on human intervention.

Food chains

This is a way of describing the transfer of energy through an ecosystem. At the start of any chain is the energy source, usually the Sun. Green plants use the Sun's energy to make their own food (by photosynthesis). Green plants and other living organisms able to use energy from the Sun in this way are termed '**primary producers**' or '**autotrophs**'; animals that eat plants get their energy by this means ('**primary consumers**', or '**herbivores**'); and other animals can obtain their energy by eating the plant-eating animal ('**secondary consumers**', or '**carnivores**'). Another animal could eat the next animal along the chain, and so on. Other animals along the chain would be described as tertiary consumers.

The relationship is known as a **food chain**. There are rarely more than four or five living things in a food chain. Note that the arrows show the direction of the energy flow, for example:

Sun → grass → rabbit → fox

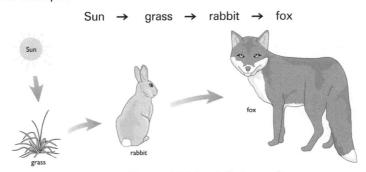

Figure 5.1 A food chain

There is a range of language to describe various points in a food chain. The full relationship could be described in terms of producers and consumers. For example, in the food chain, Sun → grass → rabbit → fox, grass is the primary producer, rabbit is the primary consumer and fox is the secondary consumer. Sometimes the chain goes beyond the secondary consumer.

As you can see, some animals can be consumers at different points along the food chain. However, animals cannot be primary producers, only plants can.

Another way of describing the relationship between *consumers* in a food chain is in terms of **predator** and **prey**. A predator is an animal that hunts and kills another animal for food. Prey is the resource for the predator. Prey always comes before predator in a food chain and energy is always transferred from prey to predator. If the population of predators is large, then the population of prey tends to decrease (because it is consumed). When the population of prey decreases, this can result in a similar decrease in the population of predators (through starvation). It can also result in an increase in the population of the plants consumed, because of the decrease in primary consumers. Thus, changes in populations at one level of a food chain will have an impact on all other levels. They will also have an impact on other food chains which have organisms in common.

Relationships can be explored in terms of the types of consumer and the versatility of a consumer to occupy various points in a food chain. Herbivores, animals that get their energy from plants, are usually primary consumers. Carnivores, animals that get their energy from other animals, are not usually primary consumers but could be secondary or tertiary consumers, or even further along the chain. **Omnivores**, animals that get their energy from plants and animals, can occupy any point along the consumer chain.

REFLECTIVE TASK

Interesting debates can arise about moral decisions and choices, e.g. for an omnivore to choose to be a primary consumer (a human who chooses to be vegetarian), as well as debates about humans providing foodstuffs for herbivores (primary consumers) which contain animal matter. What are your views on these issues?

Decomposers, i.e. living things that get their energy from dead and decaying matter, occur at all stages of the food chain. Decomposers, including fungi, maggots and bacteria, get their energy by chemically breaking down a variety of dead organisms. Animals which are decomposers include earthworms and woodlice, which obtain energy by feeding on dead and decaying organisms (plant and animal), thereby incidentally enriching soil with minerals and humus. Another term for this kind of decomposer is detrivore.

PRACTICAL TASK PRACTICAL TASK PRACTICAL TASK PRACTICAL TASK

Draw a food chain that includes bacteria. Use pictures and turn this into a mobile.

For more information on energy, see Chapter 9.

How energy efficient is a food chain?

The simple answer to this is 'not very'. Most of the light energy that falls onto a plant is reflected back; only some of it is converted into chemical energy. Some of the transferred energy is used to maintain the plant's life processes.

When primary consumers eat plants, much of the energy is used to maintain the animal's life processes. Some of the food energy is converted to other forms of energy (e.g. kinetic energy for movement) and, in the process of transfer, some of the energy is 'lost' to non-useful energy forms such as heat. Some of the food is egested and so the energy is 'lost' to the animal. Some of the energy 'lost' in a food chain, e.g. through excretion and egestion, is available to decomposers, who are themselves a part of other food chains.

When the primary consumer is eaten by secondary consumers (and so on, along the food chain), more energy is 'lost' to the system for the same reasons described above.

In general, it is estimated that 10 per cent of the energy present in primary produators is used by primary consumers for growth. Similarly, 10 per cent of the energy present in primary consumers is used for growth by secondary consumers. That is, about 90 per cent of the available energy is 'lost' at each link of a food chain. By 'lost', it is meant energy that is not useful for the next consumer.

RESEARCH SUMMARY RESEARCH SUMMARY RESEARCH SUMMARY **RESEARCH SUMMARY**

Children can have difficulty recognising the importance of the direction of the arrows in a food chain. Analysis of Key Stage 2 SATs papers in 1999 showed that many children lost marks because they drew the arrow facing the wrong way (showing 'sheep eat grass and humans eat sheep'). It is important, therefore, that children learn that the arrow represents the direction of energy flow, i.e. 'energy from the grass goes to the sheep and energy from the sheep goes to the human.' Analysis of 2005 tests indicated that children need opportunities to use information to complete food chains.

Barker and Slingsby (1998) criticised the progression of ecological concepts identified in the National Curriculum. They recommended that primary-aged children should engage in more observational work with locally occurring organisms, naming organisms and simple food chains.

Some children have difficulty understanding that plants make their food using energy from the Sun (photosynthesis), believing instead that plants get their food from the soil. Relating the origin of a food chain to a plant and, ultimately, the Sun might cause some problems (Driver *et al.*, 1994a).

Other research has shown that even secondary pupils need help in understanding that food is needed by living things as a source of energy and for growth and repair (Driver *et al.*, 1994a).

Some children think that an animal feeds on all the organisms below it in a food chain, rather than seeing a food chain as illustrating a *sequence* of feeding (Driver *et al.*, 1994b).

Younger children, especially, tend to give teleological reasons for food chains, i.e. 'there are lots of worms so that birds won't get hungry' (Leach *et al.*, 1992).

When thinking about dead organisms, children will refer to 'rotting' without knowing what is involved in the process. Few will mention the involvement of micro-organisms (e.g. bacteria or fungi) (Driver *et al.*, 1994a).

Eilam (2002) noted that the place of decomposers in a food chain is often ignored by pupils. Bacteria, in particular, are thought of more in the context of disease than as part of a food chain.

EYFS children are fascinated by slugs and snails. Bring in examples of these creatures and see what they eat. At Key Stage 1, encourage children to talk about the common plants and animals in their neighbourhood. Talk about what they eat and drink.

When first exploring food chains with Key Stage 2 children, show pictures of a range of animals and ask, 'Where does it get its food energy from?' If using pictures of dogs and cats, children are

likely to tell you their food comes from cans. You need to explore with the children where the food in the can came from so children can appreciate that the food in the cans is bits of other animals. Some will mention their pet cat getting food from the neighbour's fish pond or bringing mice and moths into the house.

Use pictures or models of familiar animals and plants and arrange them to show the direction of energy. Link the pictures with arrows. You, or a child, could write on the arrow 'gets its energy from' and/or 'is eaten by'. Pictures and models of animals and plants useful for representing common food chains can be bought commercially. Alternatively, the children could make picture cards of their own, using pictures from magazines. Hillcox (2006) offers further suggestions.

Food webs

A **habitat** is a localised environment in which organisms live and which provides all (or almost all) of an organism's needs. In a given habitat, say woodland, there is a variety of food chains, many of which interrelate. For instance, rabbits are not the only animals that get their energy from grass and grass is not their only source of energy. Foxes get their energy from animals other than rabbits. If we were to plot all the possible food chains in a given habitat, then we would find a complex set of interdependent chains. A **food web** is a way of illustrating some of this complexity and is more realistic than a food chain. Food chains are an oversimplification of the relationships within a habitat but are a useful way to introduce this aspect.

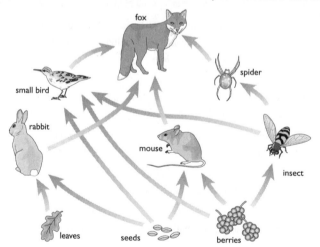

Figure 5.2. A food web

Food webs are more complex than food chains and are often incomplete because not all the feeding habits of every living thing in the web are known. Further, **parasites** would also need to be included in the web.

PRACTICAL TASK PRACTICAL TASK PRACTICAL TASK PRACTICAL TASK

Children could combine a range of food chains from the same habitat in order to show the interrelationship of living things.

You could play the food web game with your class. You will need enough labels of living things for each child in the class (including 'the Sun') and lengths of wool or string (or a ball of wool or string).

- **Organise the children into a circle.**
- **Each child represents a specific living thing and is given an appropriate label. One will represent the Sun and start by holding the wool or string.**
- **'The Sun' will start by saying something like, 'I am the Sun and my energy goes to...'**
- **The child will then select an appropriate living thing, i.e. one of the plants, and unravel the twine to reach that child.**
- **The next child might say something like, 'I am grass and my energy goes to...' Again, that child will then select an appropriate living thing, i.e. one of the animals, and unravel the twine to reach that child.**
- **This continues until the end of a food chain has been reached. The ball of twine then goes back to 'the Sun' and another food chain is started.**
- **The game finishes when all of the children have become part of the food web. All of them should have hold of the twine and you need to ensure that the twine is held tightly.**

Hillcox (2006) offers further suggestions.

REFLECTIVE TASK

It is now possible to explore the impact of particular environmental changes, e.g. the loss of a particular species of plant or animal through disease or human-imposed changes, such as the use of pesticides. If part of the food web is removed (and so the twine is released), what is the impact on other living things in the chain?

RESEARCH SUMMARY RESEARCH SUMMARY RESEARCH SUMMARY RESEARCH SUMMARY

While many children can draw and describe a simple food chain, they may have difficulty recognising interrelationships between many food chains, in the form of a food web. Many children (even at secondary school) have difficulty seeing the far-reaching effects of changes in one food chain on the whole food web and ecosystem (Driver *et al.*, 1994b).

Reiner and Eilam (2001) found that pupils, even at secondary level, tend to regard a food chain as a 'hierarchy of eating' (p564). They had difficulty understanding the role of transformation and conservation of energy and matter in the process.

When talking about feeding relationships, primary-aged children tend to focus on individual organisms (e.g. one predator and one prey) rather than relationships between populations of organisms (Leach *et al.*, 1996).

Young children at Key Stage 1 often think of living things as being dependent on human care and so could have difficulty describing a food chain that is independent of humans or an effect on a food web that is not caused by humans (Leach *et al.*, 1996).

Food pyramids

A **food pyramid** is a way of showing the quantities of energy available at each stage in a food chain. As mentioned earlier, energy consumed at one stage of a food chain does not transfer in its entirety to the next stage. Most of it is transferred to a range of work needed to keep the animal alive. Thus, one tertiary consumer will need energy from many secondary consumers and one secondary consumer will need energy from many primary consumers. This relationship can be illustrated by the concept of food pyramids (See Figure 5.3):

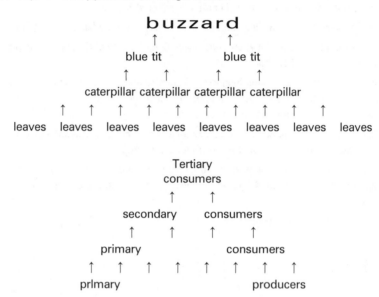

Figure 5.3 A food pyramid

Each level is described as a **'trophic' level**. As you can see, there is a decrease in the numbers (or, more accurately, the mass) of individuals at each trophic level of the food chain (though not necessarily in the mathematical proportions illustrated). In any **ecosystem**, the number (or, more accurately, the mass) of individual predators is always lower than the numbers of prey; 'mass' is more accurate because a few cabbages can support a large number of caterpillars.

Food pyramids can also be used to illustrate the effects of accumulations of toxins on a food web:

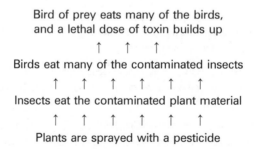

Persistent toxins become concentrated towards the top of the pyramid.

When asked to consider the impact of removing a trophic level in a food web or pyramid, children were more likely to recognise effects *up* through trophic levels than *down* through them. That is, they were more likely to recognise the effects on subsequent consumers rather than on producers, or consumers below the level removed (Leach *et al.*, 1996).

Only about 50 per cent of children from Key Stage 2 believed that producers would be the most numerous in a community. Very few children aged 5–7 believed this, opting instead for primary consumers or decomposers as the most numerous (Leach *et al.*, 1996).

Habitats and ecosystems

An ecosystem is a unit that includes all the living organisms and the physical environment. Through interactions with each other, it forms a self-sustaining system. Examples of major ecosystems include:

- tropical rainforest;
- temperate forest;
- pine forest;
- dry tropical grassland;
- desert.

Living things within an ecosystem are usually well adapted for survival in the ecosystem's particular conditions. A habitat can support only a finite number of living things and so there is competition between living things, both within a species and between different species. Habitats can be affected if part of a food web is damaged, e.g. by disease, the introduction of a new species, by deliberate intervention or by pollution. Some food chains can be destroyed; others are affected by the remaining available food source. The result is an impact on the ecosystem.

Habitat is a term used to describe the natural home of a group of plants and animals, that provides all (or nearly all) the needs of the inhabitants. Such needs include food, oxygen, water, shelter and opportunity to reproduce. The animals and plants occupying a particular habitat are called its '**community**'. The particular needs of individual populations of organisms will vary from species to species. Habitats can be of various sizes and larger habitats can have smaller habitats within them. For example, a freshwater pond can be a habitat and can be a part of a larger habitat: a woodland. Some habitats are very precise and known as micro-habitats, e.g. under a log or inside the body of an animal. 'Habitat' can be used to describe a range of environmental conditions, from the polar regions, to a desert, to a housing estate. Habitats can change through the year (because of the seasons) and this will affect the community.

Populations in a habitat can be affected by competition if more than one species feeds on a particular organism (inter-specific), one of the trophic levels is damaged, or the population of a particular organism increases in number, out of proportion to its food source (intra-specific).

Habitats can be damaged, sometimes permanently, by humans. For example, a coral reef can be damaged if too many tourists visit it. They can damage parts of the coral and reduce the food source for certain organisms. This will have a knock-on effect on other organisms. They can leave behind pollutants, causing further damage.

REFLECTIVE TASK

Think of the impact of shipping disasters and oil slicks on marine habitats. How might such habitats recover?

RESEARCH SUMMARY RESEARCH SUMMARY RESEARCH SUMMARY RESEARCH SUMMARY

Children from ages 5 to 16 can make simple links between organisms in a community, e.g. relating to food, shelter or habitat. Older children can consider abstract ideas, such as how various factors will influence interdependent populations (Leach *et al.*, 1996). Analysis of SATs tests for 2005 pupils at Key Stage 2 suggested that almost all were able to 'identify a feature of an animal that helps it to live in its environment' (QCA, 2005, p2).

Adaptation

For more information on adaptation, see Chapter 4.

Over millions of years, plants and animals will develop features that increase their chances of survival in a particular environment. That is, they become adapted to their environment. Living things that are well adapted are usually able to survive to maturity and pass on their genes to the next generation. Competition between and within species usually means that the strongest of a species will survive to pass on genes to the next generation.

Environmental issues

Consider a hedgerow

The building of new roads, extra housing and changes in farming practices have resulted in the destruction of many hedgerows. This will affect the quantities of different foods and thus have a knock-on effect up and down the food chain. Pesticides can contaminate fruits and seeds. In turn, animals can be contaminated through eating the affected food and, up the food chain, other animals will be affected by eating contaminated animals. The concentration of the pesticide increases as you move up the food chain. Using pesticides to remove 'pests' unwanted by the gardener means that a range of other wildlife has lost a potential food source. Some gardeners encourage 'top predators' into the garden as a way of controlling pests naturally. Discarded waste, such as plastic bags and cans, can damage animals and plants. Cutting hedges at nesting time can limit the availability of food along the food chain. Some people try to time trimming more carefully as well as provide nesting boxes.

Consider a river

Chemicals from farming, industry and domestic use (e.g. detergents) can initially increase the level of nutrients in the water which will be used by algae. When the food supply goes, the algae dies and rots. The result is a decrease in the level of oxygen in the water and so the fish and other animals die too. Discarded waste, such as fishing line, can harm particular birds and other animals.

A SUMMARY OF **KEY POINTS**

> Living things depend on each other for their own survival.

> The relationship between living things can be described in a number of ways, including food chains, food webs and food pyramids.

> All these terms describe the relationship in terms of energy flow.

> A food chain is not very energy efficient.

> Interference at one trophic level will have an impact on all other trophic levels.

> Habitats are a useful resource for studying the interdependence of living things.

> Over time, living things can become adapted to their environment.

M-LEVEL EXTENSION > > > > M-LEVEL EXTENSION > > > >

Reflecting on the summaries of research into how children's understanding of scientific concepts develop that you have read so far, in this and previous chapters, think about whether the media coverage of environmental issues and habitats, such as the reducing levels of rainforest and the debates over global warming, and of recent natural disasters, such as bush fires, tsunamis and earthquakes, will support or hinder the development of children's understanding of ecosystems.

REFERENCES REFERENCES **REFERENCES** REFERENCES REFERENCES

Barker, S. and Slingsby, D. (1998) From nature table to niche: curriculum progression in ecological concepts. *International Journal of Science Education*, 20(4), 479–86.

DfES (2007) *The Early Years Foundation Stage*. Nottingham: DfES Publications.

Driver, R., Squires, A., Rushworth, P. and Wood-Robinson, V. (1994a) *Making Sense of Secondary Science: Research into Children's Ideas*. London: Routledge.

Driver, R., Squires, A., Rushworth, P. and Wood-Robinson, V. (1994b) *Making Sense of Secondary Science: Support Materials for Teachers*. London: Routledge.

Eilam, B. (2002) Strata of comprehending ecology: Looking through a prism of feeding relationships. *Science Education*, 96(5), 645–71.

Hillcox, S. (2006) The survival game: teaching ecology through role play. *School Science Review*, 87(320), 75–81.

Leach, J., Driver, R., Scott, P. and Wood-Robinson, C. (1992) *Progression in Conceptual Understanding of Ecological Concepts by Pupils age 5–16*. Leeds: Centre for Studies in Science and Mathematics Education, University of Leeds.

Leach, J., Driver, R., Scott, P. and Wood-Robinson, C. (1996) Children's ideas about ecology 3: ideas found in children aged 5–16 about the interdependency of organisms. *International Journal of Science Education*, 18(2), 129–41.

QCA (2005) *Implications for teaching and learning from the 2005 national curriculum tests*. London: QCA.

Reiner, M. and Eilam, B. (2001) Conceptual classroom environment: A system view of learning. *International Journal of Science Education*, 23(6), 551–68.

FURTHER READING FURTHER READING **FURTHER READING** FURTHER READING

Centre for Life Studies (no date) *A Freshwater Food Chain Mobile*. London: Zoological Gardens.

de Boo, M. (2000) *Laying the foundations in the early years*. Hatfield: Association for Science Education.

DfE (2011) *Teachers' Standards*. Available at www.education.gov.uk/publications.

Nuffield Primary Science (1995) *Living Things in their Environment: Teachers' Guide*, 2nd edn. London: Collins Educational.

6
Materials

Curriculum context

National Curriculum programmes of study

At Key Stage 1, children should be taught about how to use their senses to explore and recognise similarities and differences between materials, to sort objects into groups on the basis of simple properties, to recognise and name common types of material and recognise that some are found naturally, to find out about the uses of a variety of materials and how these are chosen for specific uses on the basis of their simple properties, and to find out how the shapes of objects made from some materials can be changed by some processes, including squashing, bending, twisting and stretching.

At Key Stage 2, children should be taught to compare everyday materials and objects on the basis of their material properties, including hardness, strength, flexibility and magnetic behaviour, and to relate these properties to everyday uses of materials, that some materials are better thermal insulators than others, that some materials are better electrical conductors than others, to describe and group rocks and soils on the basis of their characteristics, including appearance, texture and permeability, to recognise differences between solids, liquids and gases in terms of ease of flow and maintenance of shape and volume, to describe changes that occur when materials are heated or cooled, that temperature is a measure of how hot or cold things are, about reversible changes, including dissolving, melting, boiling, condensing, freezing and evaporating, and the part played by evaporation and condensation in the water cycle.

Early Years Foundation Stage

In developing their Knowledge and Understanding of the World young children should: notice and comment on patterns; show an awareness of change; explain their own knowledge and understanding; and ask appropriate questions of others. The Early Learning Goals suggest that young children should: investigate objects and materials by using all their senses as appropriate; find out about and identify some features of living things, objects and events they observe; look closely at similarities, differences, patterns and change; ask questions about how things happen and how things work.

Introduction

All materials are made up of a combination of one or more of over 100 elements. The resultant variety is staggering. Materials can be transparent, brittle, toxic, soft, good conductors of electricity, and so on. The same material can behave differently when it exists as a solid, liquid or gas (try crunching liquid water!). Some materials can react with others to produce new materials and by-products. Without materials, life just would not be.

What are materials made of?

Knowing about the atomic structure of a substance is the key to understanding its properties and how it will behave in particular circumstances. This chapter will be looking at ways of describing materials (substances) at a level that cannot be seen with the naked eye. Much of the chapter will help you as a teacher in your understanding. However, any explanation using this information will be beyond the understanding of most of the children you teach. What you should have is a better awareness of children's conceptual struggles. An understanding of the structure of materials has been developed by scientists based on the behaviour of materials under different conditions.

You may remember learning about the periodic table in secondary school. There are over 100 (between 111 and 118 depending on your source, 92 of them occurring naturally) members of the periodic table and they are known as **elements**. All materials in the world are made up of these elements, either singly or in some form of combination. Each element has been given its own symbol, e.g. O for oxygen, H for hydrogen, Pb for lead (Latin is *plumbum*), Fe for iron (Latin is *ferrum*).

In the nineteenth century, Dalton tried to explain the behaviour of certain chemical reactions by imagining that all matter was made up of **atoms**, small, hard and indivisible spheres. Today, we have evidence that atoms are made up of even smaller particles and it is these smaller particles that will be described. When considering particles and their make-up, it is necessary to stretch the imagination beyond what can easily be seen, even with a school microscope. With powerful electron microscopes, which magnify up to two million times, large single atoms or groups of atoms can be seen.

Elements

An element is the smallest unit that cannot be divided into simpler substances by normal chemical means. Atoms of the same element are identical and have the same **atomic number** (that is, the same number of protons) but, because of the existence of isotopes, not always the same **atomic mass** (the total number of protons and neutrons). **Isotopes** of elements differ in the number of neutrons, but their chemical properties are identical.

Atoms

For more information on electricity, see Chapter 8.

An atom is a single unit of any element. It is the smallest particle that can be part of a chemical reaction. An atom is made up of **protons, neutrons** and **electrons**. The nucleus (containing protons and neutrons) is at the core with electrons occupying the surrounding layers. There are other, smaller, units that make up an atom (for example, quarks) but the three basic components mentioned will be considered here. Atoms are of different sizes. The smallest complete atom is hydrogen. Atoms are electrically neutral, i.e. they have no charge. This is because the number of protons in the nucleus balances the number of electrons occupying the shells surrounding the nucleus. Most of the atom is empty space.

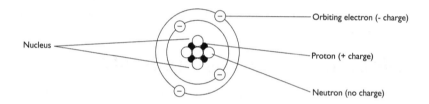

Figure 6.1 Representation of an atom

Protons, neutrons and electrons

Protons are particles which make up the nucleus of an atom. They have a positive charge and are much more massive than electrons but the unit of positive charge is equal to the negative charge of an electron. One proton is described as having an atomic mass of 1. The number of protons in an element is equal to that element's atomic number.

Neutrons are particles which make up the nucleus of an atom. They have the same mass as a proton but no charge and are there so that the positively charged protons will stay together inside the nucleus (remember, 'like charges' repel each other). They act like glue, holding the protons in the nucleus. Isotopes of elements differ in the number of neutrons, but their chemical properties are identical.

Electrons orbit the nucleus of an atom. They each have a negative charge of differing quantities. They behave as if they occupy different layers or 'shells', or energy levels. Electrons with the same quantity of energy are on the same energy level. They are attracted to the protons in the nucleus, although the further they are from the nucleus, the easier it is for them to leave the atom and the more energy they have. In some elements, it is easier for the electrons to be removed than in others. As a general principle, electrons behave as if there is a maximum of two occupying the shell (or energy level) nearest to the nucleus. The second and third shells can each contain a maximum of eight electrons. The innermost shells are filled with electrons before the next shell is occupied. It is the electron structure and distribution of electrons that is responsible for an atom's chemical properties. In some materials, some electrons are easily removed and this accounts for certain physical properties such as electrical conductivity. In a neutral atom, the number of electrons matches the number of protons. If electrons are removed, then the atom becomes positively charged (because there are now more protons than electrons) and is known as a positive **ion**. If electrons are gained, then the result is a negative ion (because there are now more electrons than protons).

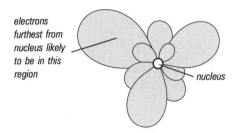

electrons furthest from nucleus likely to be in this region — nucleus

An alternative model of an atom, which better represents the behaviour of electrons, portrays them as occupying fuzzy-edged balloon-shaped regions. Electrons furthest from the nucleus (with the highest energy levels) occupy the largest 'balloons'.

Isotopes

Some elements consist of atoms of the same atomic number but different mass numbers. This is because the number of neutrons in the nucleus can vary. For example, all hydrogen atoms contain one proton in the nucleus and one electron in the first shell (or energy level). Most have no neutron. However, a very small minority have, in addition to the one proton, one neutron in the nucleus, and an even smaller percentage have two neutrons. So, even though the atomic number for each isotope of hydrogen is 1, the atomic mass can have values from 1 to 3.

Name of element	Symbol of element	Number of protons	Electrons in first shell	Electrons in second shell	Electrons in third shell	Electrons in fourth shell	Group number
Hydrogen	H	1	1				1
Helium	He	2	2				0
Lithium	Li	3	2	1			1
Beryllium	Be	4	2	2			2
Boron	B	5	2	3			3
Carbon	C	6	2	4			4
Nitrogen	N	7	2	5			5
Oxygen	O	8	2	6			6
Fluorine	F	9	2	7			7
Neon	Ne	10	2	8			0
Sodium	Na	11	2	8	1		1
Magnesium	Mg	12	2	8	2		2
Aluminium	Al	13	2	8	3		3
Silicon	Si	14	2	8	4		4
Phosphorus	P	15	2	8	5		5
Sulphur	S	16	2	8	6		6
Chlorine	Cl	17	2	8	7		7
Argon	Ar	18	2	8	8		0
Potassium	K	19	2	8	8	1	1
Calcium	Ca	20	2	8	8	2	2

Table 6.1

How does this relate to elements in the periodic table?

The periodic table was developed in the nineteenth century by the Russian, Dmitri Mendeleev. It was a way of ordering elements by their atomic mass (now they are ordered by atomic number) but also of making sense of patterns of behaviour of particular elements. 'Families' of elements were arranged in rows. As new elements were discovered or made, they were added to the periodic table, continuing the patterns identified. It is possible to link the properties of elements to where they are placed in the periodic table. For an excellent source about the periodic table go to www.webelements.com/

REFLECTIVE TASK

Using the information in Table 6.1 and a highlighter pen, identify elements with the same number of electrons in their outermost shell. Use this to compare where they are placed in the periodic table.

Patterns can be found in the periodic table for the number of electrons, neutrons and protons each element has. Thus, the number of electrons in the outer energy shell is the same as the group number in the periodic table (with the exception of the noble **gases**, whose outer electron shells are full and which are in group 0).

The atomic number of any element indicates how many protons are in the nucleus and each succeeding element has one extra proton in its nucleus.

Atomic number = number of protons

Elements in the periodic table are arranged according to their relative atomic number. Atomic mass is the sum of the number of protons and neutrons in the nucleus. It is the average of the mass numbers of all the isotopes. Since more than 99.9 per cent of the isotopes of hydrogen have an atomic mass of 1 and the rest an atomic mass of 2 or 3, the atomic mass is only just greater than 1. The atomic number is 1.

Atomic mass = number of protons + number of neutrons

For more information on exothermic and endothermic reactions, see Chapter 7.

A **compound** is a substance composed of two or more elements in definite proportions by weight, e.g. carbon monoxide (CO) and sulphuric acid (H_2SO_4), which are bonded together and cannot be separated by physical means (although they can be separated in a chemical reaction). Compounds are held together by ionic or covalent bonds. These bonds contain energy, bond energy, and this differs for different compounds or **molecules**. In a chemical reaction, these bonds are broken and new bonds are formed because new chemicals are formed. The energy needed to form the new bonds can differ in quantity from the bond energy in the original chemicals.

A compound is different from a **mixture**, which consists of different substances (elements and compounds) mixed together. The substances are not bonded to each other and there is no chemical reaction between them. Mixtures are usually easy to separate by physical means. Most natural forms of material on the Earth are mixtures, e.g. air is mainly a mixture of nitrogen gas and oxygen gas, with some other gases.

RESEARCH SUMMARY RESEARCH SUMMARY RESEARCH SUMMARY **RESEARCH SUMMARY**

Ahtee and Varjola (1998) cited a range of research to support the assertion that pupils, even at Key Stage 4, rarely use terms such as atoms and molecules when explaining chemical reactions and do not easily distinguish between such terms as atoms, elements, molecules or compounds. This is in spite of an apparent understanding of such terms previously in a different context. Piaget and Inhelder's work (1974) described two main misconceptions.

- Atomistic, that is, that matter is made up from particles whose properties resemble those of atoms.
- That matter is continuous, that is, it isn't made of discrete particles, even though pupils accept a particle model in a very limited context.

They went on to say that the research showed that children are unlikely to accept the concept that molecules of air are moving in an empty space. Johnson (2002) states that pupils need to have a basic particle model before they can begin to understand the ideas of atoms and bonding.

Skamp (2005) argues that direct teaching and 'scaffolded' teaching about particles might lead to some conceptual understanding but it is highly context-dependent and so rarely applied to a new context. However, primary teachers should raise awareness to provide a grounding for later learning, focusing on what is observable.

PRACTICAL TASK PRACTICAL TASK **PRACTICAL TASK** PRACTICAL TASK

For young children, provide a few (up to 6) common materials to explore, and ask questions such as: Can you see through it? What sound does it make when you tap it? Will it bend? For older children, try exploring similar materials and raising questions such as: Do all woods float? Are some more easily scratched? Do all metals conduct electricity? Are they all attracted to a magnet? Try to ensure that a range is used. Make a note of the questions the children want to explore, and of the explanations and observations they give.

A molecule is a group of two or more atoms bonded together. These can be atoms of the same element or of two or more different elements. They vary in size and complexity, e.g. a hydrogen molecule (H_2) compared with a sugar molecule ($C_6H_{12}O_6$). The latter is the symbolic representation of a sugar molecule and is known as its **chemical formula**. The subscript numbers tell you how many atoms of each element there are in one molecule. So, one molecule of sugar has 6 atoms of carbon, 12 atoms of hydrogen and 6 atoms of oxygen.

When atoms join, they bond together. There are two main types: **ionic bonding** and **covalent bonding**. The kind of bond between atoms depends on the type of atoms.

Ionic bonding

This form of bonding is common when a metal atom combines with a non-metal atom, for example when sodium combines with chlorine to form sodium chloride (common table salt – see Figure 6.2.)

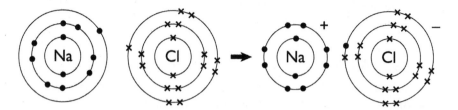

**Figure 6.2 Dot and cross diagram to show ionic bonding
between sodium and chlorine**

Atoms will tend towards achieving a stable electronic arrangement. That is, they will lose or gain electrons in order to be stable, like the noble gases (e.g. helium, argon). The electrons move around the nucleus at certain energy levels (or in certain shells) and there is a maximum number of electrons that can be accommodated at each energy level. A noble gas has the maximum number of electrons at each energy level. Other elements will have less than the maximum number, to varying degrees.

A sodium atom has one electron more than the noble gas neon and so will readily lose this electron. A chlorine atom has one electron fewer than the noble gas argon and so will readily accept an electron. If a sodium atom loses an electron and a chlorine atom gains an electron, then both will have achieved a stable electronic configuration.

If sodium loses an electron, then there will be an excess number of protons against the total number of electrons. Thus, the sodium atom will no longer be electrically balanced, but will have a positive charge. When an atom is positively or negatively charged, we call this an ion. The sodium ion is represented by Na^+ and now has an electron arrangement of 2,8 instead of 2,8,1. Other metals, like sodium, tend to form positive ions. Metals in the same periodic group as sodium form ions with one excess proton (K^+). Metals in the neighbouring group form ions with two excess protons (Mg^{2+}).

Similarly, if chlorine gains an electron, then there will be an excess number of electrons against the total number of protons. Thus, the chlorine will no longer be electrically balanced, but will have a negative charge. The chlorine ion is represented by Cl^- and now has an electron arrangement of 2,8,8 instead of 2,8,7. Other non-metals, like chlorine, tend to form negative ions. Non-metals in the same periodic group as chlorine form ions with one excess electron (Fl^-). Non-metals in the neighbouring group form ions with two excess electrons (S^{2-}).

The positively charged sodium ions and negatively charged chlorine ions are strongly attracted to each other. When millions of these charged ions come into contact, they will hold on to each other. The sodium and chlorine ions are held together, in a regular, lattice arrangement, by strong electrostatic forces. The crystal structure of sodium chloride is a lattice arrangement of cubes, with each chlorine ion surrounded by six sodium ions and vice versa. The forces between the ions in a crystal of sodium chloride are very strong and so a lot of energy is needed to break them. Hence, the **temperature** at which sodium chloride will melt is very high. Similarly, the boiling point is also very high. We can generalise the properties of materials as a result of ionic bonding.

- In solid form they are crystalline in structure (they are usually solid at room temperature).
- The strong bonds mean that the melting point is high.
- In solid form, they don't usually conduct electricity (because the charged ions are in fixed positions and are not free to move about).
- When they dissolve in water, the solution will conduct electricity because the charged ions are able to move about.
- When they are melted, and therefore a liquid, they will conduct electricity, because the charged ions are able to move about.

REFLECTIVE TASK

Will table salt conduct electricity? If dissolved in water, it will. Can you give an explanation?

Elements in Groups 1 and 2 will tend to have ionic bonding when forming compounds with elements from Groups 6 and 7 of the periodic table.

Covalent bonding

This is a form of bonding where electrons are *shared* rather than taken or given, as in ionic bonding. It is commonly found when two non-metals combine. The purpose of covalent bonding is the same as that for ionic bonding, namely to achieve a stable electron arrangement. Thus, when electrons are shared between atoms, each atom will have a full outer shell.

If we consider two hydrogen atoms, each atom has one electron. If they join together, sharing their electrons, then each atom will have a full outer shell:

$$H + H \rightarrow H_2$$

In diagrams, a single line represents a bond of this kind: H—H.

Methane (CH_4), a gas formed in our intestines, is another example of a molecule with covalent bonding. Carbon has four electrons in its outer shell and the four electrons from four atoms of hydrogen complete the outer shell. Likewise, the four electrons from the carbon atom are used by each of the hydrogen atoms to complete their outer shells:

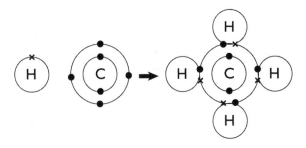

Figure 6.3 Dot and cross diagram to show the sharing of electrons in methane

It is possible to have double and triple covalent bonds, as well as the single covalent bonds described above. In double covalent bonds, each atom shares two electrons, and in triple covalent bonds, each atom shares three electrons.

Oxygen, O_2, consists of two oxygen atoms, each sharing two electrons. In diagrams, double bonds are shown as: O=O. Nitrogen molecules each contain two nitrogen atoms held together with a triple covalent bond and so each nitrogen atom is sharing three electrons. In diagrams, triple bonds are shown as: N≡N.

We can generalise the properties of small molecules that have covalent bonding.

- They tend to have low melting and boiling points; at room temperature, they are often gases or liquids and solids with low boiling points because the force holding the molecules together is relatively small.
- They are usually insoluble in water but soluble in organic solvents.
- They tend not to conduct electricity because there aren't any ions and neither are there any spare electrons moving around.

In summary then, when we look at our world, we can predict that things that look like crystals and are hard are likely to have ionic bonding, whereas soft, runny or gas-like materials are likely to consist of molecules joined together by covalent bonding.

What kind of bonding is there in water?

Water (H_2O) is runny with quite a low boiling point (100°C) which suggests that the forces *between* molecules are not very large, although *within* each molecule the forces are very large. The electron of each of the two hydrogen atoms is shared with the oxygen atom so that each atom achieves a stable electronic arrangement. Thus, the type of bond is covalent: H–O–H. However, this isn't the complete story because water is rather unusual. One of the hydrogen parts of the molecule is slightly positively charged and the other part of the molecule (the hydroxyl part, O–H) is slightly negatively charged. Thus, when salt is added to water, the slight negative charges in the water are able to attract the sodium chloride ions and pull them off the salt crystals, thus dissolving the salt. Molecules that are truly covalent (such as cooking oil) do not have any charges on the molecules and so water cannot attract molecules and pull them away. Thus, cooking oil is not soluble in water.

RESEARCH SUMMARY RESEARCH SUMMARY **RESEARCH SUMMARY RESEARCH SUMMARY**

Younger children tend to choose criteria to sort materials where they are actively involved, e.g. whether the materials can be bent or whether they produce sound. Moving towards Key Stage 2, they will use more complex criteria for sorting, such as function, so they will sort a particular object in terms of its belonging to a larger group such as construction materials or food. They tend to generalise about particular properties and the properties they will associate with metals are that they are strong, shiny and sharp (Russell *et al.* 1991).

REFLECTIVE TASK

Reflect on children's responses from the earlier practical task and try again with a sorting acitivity. Ask a group of children to sort a variety of 10 to 15 materials (fewer for younger children) and note criteria used. Repeat with a group of children from a different age group and compare the criteria used. Allow younger children plenty of time to choose their own criteria before suggesting some of your own.

Chemical reactions form new substances

In chemical reactions new substances are formed. The new substances can be very different from the original chemical reactants. For instance, sodium is a highly reactive metal and is a solid at room temperature. It has to be stored very carefully under oil because it will readily ignite on contact with oxygen in the air. Chlorine is a highly toxic substance and is a green gas at room temperature. When these two substances react with each other, they will form sodium chloride (common table salt), a harmless, crystalline solid at room temperature.

Physical changes do not result in new substances being formed

Physical changes include changes of state, i.e. from **solid** to **liquid**, solid to gas, liquid to gas or in the reverse direction. There is no chemical reaction. The mass of the material does not change, although, when liquid water becomes solid water (ice), the volume changes and so some children might think that there is more water in the 'ice' state. This is not so. The mass of water, i.e. the amount of water, does not change as it becomes ice. Although the volume increases, the density of the solid water decreases (solid water floats in liquid water) and the mass remains the same (remember that density is the ratio of mass to volume). In fact, water is at its most dense at 4°C (when it is in a liquid state). Physical changes are usually reversible, e.g. water vapour can condense and then solidify, or vice versa.

For more information on solids, liquids and gases, see below.

Conducting heat – another property of materials

If you put a metal teaspoon into a cup of hot coffee, the particles at the bowl end will absorb the **heat** energy and vibrate much faster. This will affect nearby particles, causing them to vibrate more quickly and thus passing on the energy. After a while, the handle will become hot too. We say the heat is conducted when the particle movement is handed from one particle to another.

Most of the energy in conductors is in fact carried by the 'free electrons', and only a little by vibration. Metals have one or two orbiting electrons which are so loosely bound to the nucleus that they can be easily freed and move around (rather like the freedom of movement particles have when they are in a gaseous state). It is these roaming electrons that absorb energy from the source of heat (e.g. the cup of hot coffee).

Metals are the best conductors of heat. Most non-metals (e.g. woods and plastics) are poor conductors and so are most liquids. Gases are the worst conductors.

In non-metals the energy is transferred by the handing on of increased vibration but, because there are no free electrons to pass on the energy, the only way that heat can travel along the material is by direct transfer of energy from one atom to another.

Materials (solid, liquid or gas) which are poor conductors of heat are known as insulators. Many materials are good insulators because they have tiny pockets of air trapped in them (e.g. duvets and padded jackets). Air is a poor conductor of heat, but a good insulator.

Touch can be used to sort some conductors and insulators. Metal feels cold because body heat is conducted away from your hand. Carpets feel warm because they hardly conduct away any heat from the soles of your feet.

PRACTICAL TASK PRACTICAL TASK **PRACTICAL TASK** PRACTICAL TASK

Ask children to sort a range of materials into those that are good at taking heat away from their hands and those that are not as good. Which materials are best for gloves? Which are best for cooling the skin?

What is a solid, a liquid and a gas?

To explain this, it is helpful to appreciate that all things are made up of particles (molecules and atoms). It is the way these particles *behave* which determines whether something is a solid, liquid or gas (at a particular temperature and atmospheric pressure).

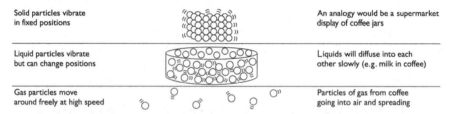

Solid particles vibrate in fixed positions		An analogy would be a supermarket display of coffee jars
Liquid particles vibrate but can change positions		Liquids will diffuse into each other slowly (e.g. milk in coffee)
Gas particles move around freely at high speed		Particles of gas from coffee going into air and spreading

Figure 6.4 Particulate level of solids, liquids and gases

Each particle is always moving (vibrating). If they are very close together then they may be attracted to each other and stick together. In a solid, the particles are strongly attracted to each other and are held together (see Figure 6.4). Although particles can't move to another place, they will vibrate about their fixed positions. We 'see' solids as having a fixed shape and volume and being impossible to compress. Two solids generally won't diffuse into each other (a coffee bean and a tea leaf).

In a liquid, the particles are still very close together but their vibrations are so vigorous that the forces of attraction are weaker and therefore can't hold them in fixed positions. They roll around each other. We 'see' liquids as able to flow into any shape but having a fixed volume and, like solids, they cannot be compressed.

In a gas, the particles move about in all directions at high speeds, colliding with each other as well as with the walls of the container. There's hardly any attraction between particles and they're too spread out to stick together. Gases have no fixed shape and no fixed volume. A gas will fill any space available. A gas can be compressed into a smaller space. Gases will quickly diffuse into each other.

RESEARCH SUMMARY RESEARCH SUMMARY **RESEARCH SUMMARY RESEARCH SUMMARY**

Children's misconceptions include the following:

- particles are different according to their state, i.e. gas molecules are round, molecules of liquid have an irregular form and molecules of solids are cubic or cuboid (Haider and Abraham, 1991);
- the size of molecules is determined by the state of matter, i.e. molecules are largest in solids and smallest in gases (Pereira and Pestana, 1991).

Solids are identified as: hard, can be held or touched, heavier than liquids. Solids can often be too different from each other for children to be able to develop generalising criteria for classification.

Liquids are identified as: can flow and can be poured, they run and are heavier than gases. They are easier to classify than solids. Children tend to compare liquids in terms of how similar they are to water. Liquids that are coloured or much more viscous than water, e.g. treacle, are more difficult to classify as a liquid.

Gases are more difficult for children. They tend to ascribe liquid properties to gases, i.e. poured or occupies a certain volume, weightless. Air is often described by children as one gas and it is seen as a different substance from other gases. Sprays and mists, smoke and flames are regarded as gases – gas is the particular fuel in their house – it is dangerous and intangible. They tend to equate gas, air and oxygen as all meaning the same thing.

Children find pastes, foams, powders and any soft substance difficult to classify. They might classify them as being something between a liquid and a solid. The bigger the particle, then the more easily children classify this as a solid. Particle here means pieces of solid, not particles at the atomic level. Ice is sometimes not classified as a solid because it can be changed into water. Children often believe that the weight increases when a liquid changes to a solid state and that it decreases when it changes to a gaseous state (Krnel *et al.*, 1998).

Krnel *et al.* (2005) found that by the age of 13 pupils tended to consider five states of matter (rather than the accepted three), namely: hard rigid solids, pliable solids, powders, liquids and gases. Young children had an 'undifferentiated "object matter" conception' (p377), tending just to name substances or objects. They suggest helping pupils to become familiar with a range of behaviours of common substances ('prototypes') and then to use these to make comparisons and build up experiences with less familiar materials. Understanding can also be developed by helping and encouraging pupils to move 'from the action language of verbs to the descriptive language of adjectives' (p381) so that the properties of materials are described more objectively.

Johnson (1998) summarises five important aspects of particle theory.

- Children tend to show spacing in a liquid as being somewhere between the spacing between solids and gases but they tend to underestimate the spacing for the gaseous state.
- Children show very little appreciation of the intrinsic motion of the particles.
- Very few children use ideas of forces or attraction or cohesion between particles, even when considering solids, or particles in a solid state and the space between particles.
- Children have great difficulty with the idea of there being nothing between particles, especially when they consider gases. They prefer to consider that there is something between the particles and they usually use the term air (whatever they mean by this term).
- They tend to attribute microscopic properties to individual particles. 'Pupils are not seeing the properties of state in terms of the collective behaviour of particles.'

Is denim a solid, liquid or gas? How would you explain your answer to a child?

How do solids, liquids and gases expand?

Solids, liquids and gases all expand when heated and contract when they are cooled. Water heated from 0° to 4°C is an exception to this and so is wood, which contracts (mostly because of chemical changes) when it is heated.

Usually, the change in size is too small to notice, although many of us have seen (or seen pictures of) railway tracks which have buckled when the weather has been unusually hot, because the metal has expanded.

With solids, when they are heated, the particles vibrate more vigorously. These vibrations take up more space and so the particles move further apart. On a larger scale, this results in a heated solid expanding. Some solids expand more than others, e.g. if a 1m bar of aluminium is heated to 100°C, it will expand by 3mm, whereas a 1m bar of glass will expand by 0.9mm and a 1m bar of Pyrex by 0.3mm. This can be put to good use in cooking. When a glass dish is put into a hot oven, the outside of the dish expands before the inside. This creates a strain which will crack the dish. Pyrex doesn't expand as much as glass and so is unlikely to crack.

Liquids expand when heated for the same reasons as solids. This can be put to good use in thermometers where the liquid is given space to expand (or contract) along a thin tube when the bulb of the thermometer is touching something. Liquids expand much more than solids.

Gases expand more than liquids or solids. Since gases can be squashed (compressed) it is possible to contain and stop a gas from expanding. However, if it is stopped, then the gas will push against the container and the pressure inside the container rises. If the container is not strong enough, then the gas will force its way through it. This is why aerosol cans should never be left on windowsills. Heat from the Sun can cause the gas inside the can to expand so much that the can will explode.

Changes of state

For more information on energy, see Chapter 9.

Before considering how changes of state can be brought about by transferring energy, we need to consider a distinction that causes confusion for children and adults alike namely, what is heat ... what is temperature?

- **Heat** is a form of kinetic energy (kinetic energy is any type of movement energy). In everyday language, 'heat' and 'temperature' are often used to mean the same thing. To scientists, they mean quite separate things. When heat is given to something, e.g. a kettle of water supplied with heat from an electric element (electrical energy transferred into heat energy), then the temperature of the water increases. Heat is the *cause* and a rise in temperature is the *effect*.
- **Temperature** is the degree of hotness of something. It is possible for something to be at a very high temperature and yet have very little heat. For instance, the spark from a sparkler is a white-hot fragment of iron and yet if it lands on your hand it doesn't give out enough heat on cooling to

burn your hand. Compare this with the relatively low temperature of a teaspoon of boiling water. Now imagine the kind of burn a teaspoon of boiling water would give your hand, because there is a relatively high amount of heat energy.

Scientists have a general hypothesis that when two bodies at different temperatures are placed in contact (e.g. your hand and a spark from a sparkler), heat energy (also known as thermal energy) will always flow from the one at the higher temperature to the one at the lower temperature (regardless of relative size). This flow of heat continues until the temperature of both 'bodies' is the same.

At an atomic level remember that particles are always moving. Heat is the energy of movement: the kinetic energy of particles. At the level of an object, the total heat is the sum of the kinetic energies of all the molecules. In the everyday world, all particles move and so this means all objects made of particles – e.g. ice, sand or lava – have heat energy. Scientists have managed to cool something until the particles have almost stopped moving and cannot go any slower. The temperature at which this would be achieved is –273°C (absolute zero).

Temperature is a measure of the average kinetic energy of the particles, i.e. how fast the particles are moving, on average. When the temperature falls, particles move more slowly, and when the temperature rises, particles move more quickly. If two objects have the same temperature then this means their particles have the same average kinetic energy.

So, if two objects with different temperatures come into contact with each other, then the heat energy will move from the object whose particles are moving fastest until the particles of both objects are moving as fast as each other, on average.

Because heat is a form of energy, it is measured using the same unit as all other forms of energy, the joule (J). Temperature can be measured using a thermometer and the scale used for measuring varies. The thermometers for everyday living measure temperature in degrees Celsius (°C). Celsius is the surname of the Swedish astronomer who devised the scale over 200 years ago. The two fixed points are 0°C and 100°C. These represent the freezing point and boiling point of *pure* water at normal atmospheric pressure. Absolute zero (see above) is 0K on a scale named after Kelvin (a Scottish physicist) who devised it about 100 years ago. 0°C is equivalent to 273K and 100°C is equivalent to 373K.

RESEARCH SUMMARY RESEARCH SUMMARY RESEARCH SUMMARY **RESEARCH SUMMARY**

Papageorgiou and Johnson (2005) found some evidence to suggest that introducing ideas about particles to a small sample of primary children helped their understanding of changes of state and mixing. Further, there appears to be conceptual progression from: 'particles' has no meaning → particles being additional to the substance → particles are the substance but are just small bits of the substance → science acceptable explanations using particle ideas.

PRACTICAL TASK PRACTICAL TASK **PRACTICAL TASK** PRACTICAL TASK

Put a dented table tennis ball in very hot water. Explain how this helps to remove dents.

A SUMMARY OF KEY POINTS

> All materials are made up of elements joined together in a wide range of combinations.

> The periodic table identifies such elements and represents them in increasing atomic number, grouping similar elements into 'families'.

> Atoms can join together in different ways, namely by ionic or covalent bonding.

> The kind of bonding can influence the properties of a material.

> Chemical reactions result in the formation of new substances, which can be very different from the original chemical reactants.

> Physical changes, including changes of state, do not result in the formation of new substances.

> Materials can exist in different states, i.e. as a solid, liquid or gas, although their melting points and boiling points will differ greatly from each other.

> For a particular material to change state, energy is required.

M-LEVEL EXTENSION > > > > M-LEVEL EXTENSION > > > >

As you will have seen from the tasks suggested in this chapter, there are many practical activities that can be planned to develop children's understanding of materials, but what investigations or experiments can be carried out in the primary classroom to help to prevent the common misconceptions identified in the research cited in the summaries provided? Discuss your ideas with your fellow trainees, class teacher colleagues and an experienced science subject leader.

REFERENCES REFERENCES **REFERENCES** REFERENCES REFERENCES

Ahtee, M. and Varjola, I. (1998) Students' understanding of chemical reactions. *International Journal of Science Education*, 20(3), 305–16.

Haider, A. H. and Abraham, M. R. (1991) A comparison of applied and theoretical knowledge of concepts based on the particulate nature of matter. *Journal of Research in Science Teaching*, 28(10), 919–38.

Johnson, P. (1998) Progression in children's understanding of a 'basic' particle theory: a longitudinal study. *International Journal of Science Education*, 20(4), 383–412.

Johnson, P. (2002) Children's Understanding of Substances. Part 2: explaining chemical change. *International Journal of Science Education*, 24(10), 1037–54.

Krnel, D., Watson, R. and Glažar, S. A. (1998) Survey of research related to the development of the concept of matter. *International Journal of Science Education*, 20(3), 257–89.

Krnel, D., Watson, R. and Glažar, S. A. (2005) The development of the concept of 'matter': a cross-age study of how children describe materials. *International Journal of Science Education*, 27(3), 367–83.

Papageorgiou, G. and Johnson P. (2005) Do particle ideas help or hinder pupils' understanding of phenomena? *International Journal of Science Education*, 27(11), 1299–317.

Pereira, M. P. and Pestana, M. E. (1991) Pupils' representations of models of water. *International Journal of Science Education*, 13(30), 313–19.

Piaget, J. and Inhelder, B. (1974) *The Child's Construction of Quantities*. London: Routledge & Kegan Paul.

Russell, T., Longden, K. and McGuigan, L. (1991) *SPACE – Materials*. Liverpool: Liverpool University Press.

Skamp, K. (2005) Teaching about 'stuff'. *Primary Science Review*, 89, 20–22.

DfES (2007) *The Early Years Foundation Stage*. Nottingham: DfES Publications.

FURTHER READING FURTHER READING **FURTHER READING** FURTHER READING

Archer, D. (1991) *What's Your Reaction?* London: The Royal Society of Chemistry.

Bell, D. and George, N. (eds) (1997) *Nuffield Primary Science – Understanding Science Ideas: a Guide for Primary Teachers*. London: Collins Educational.

Bethell, A., Dexter, J. and Griffiths, M. (1996) *Heinemann Co-ordinated Science: Chemistry*. Oxford: Heinemann Educational & Professional Publishing Ltd. Any other good textbook for GCSE science should enable you to extend your knowledge.

de Boo, M. (2000) *Laying the Foundations in the Early Years*. Hatfield: Association for Science Education.

DfE (2011) *Teachers' Standards*. Available at www.education.gov.uk/publications.

Nuffield Primary Science (1995) *Materials – Teacher's Guide*, 2nd edn. London: Collins Educational.

Pendlington, S., Palacio, D. and Summers, M. (1993) *Understanding Materials and Why They Change*. Oxford: Oxford University Department of Educational Studies and Westminster College.

Taber, K. (2002) *Chemical Misconceptions – Prevention, Diagnosis and Cure. Volume 1: Theoretical Background*. London: Royal Society of Chemistry.

7
Particle theory and the conservation of mass

Curriculum context

National Curriculum programmes of study

At Key Stage 1, children should be taught how to explore and describe the way some everyday materials change when they are heated or cooled.

At Key Stage 2, children should be taught to describe and group rocks and soils on the basis of their characteristics, including appearance, texture and permeability, how to describe changes that occur when materials are mixed, describe changes that occur when materials are heated or cooled, about reversible changes, including dissolving, melting, boiling, condensing, freezing and evaporating, about the part played by evaporation and condensation in the water cycle, that non-reversible changes result in the formation of new materials that may be useful, that burning materials results in the formation of new materials and that this change is not usually reversible, how to separate solid particles of different sizes by sieving, that some solids dissolve in water to give solutions but some do not, how to separate insoluble solids from liquids by filtering, how to recover dissolved solids by evaporating the liquid from the solution and to use their knowledge of solids, liquids and gases to decide how mixtures might be separated.

Early Years Foundation Stage

In developing their Knowledge and Understanding of the World young children should: notice and comment on patterns; show an awareness of change; explain their own knowledge and understanding; and ask appropriate questions of others. The Early Learning Goals suggest that young children should: investigate objects and material by using all their senses as appropriate; find out about, and identify, some features of living things, objects and events they observe; look closely at similarities, differences, patterns and change; ask questions about how things happen and how things work.

Introduction

The way materials behave can usually be explained at the level of particles, i.e. atoms and molecules. At this level, particles cannot be seen with the naked eye. Teachers are asking children to think at an abstract level, about behaviours which they cannot see directly. For this reason, many have difficulty with their understanding.

Dissolving: what does it mean?

What happens when you add sugar to water? It dissolves. What does this mean? Each crystal of sugar is made up of millions of molecules of sugar (containing the atoms of hydrogen, oxygen and carbon). These molecules are held together strongly to make up the crystal. The molecules of water are moving about and frequently bump into the molecules that make up the sugar crystals. Sometimes they pull away some of the sugar molecules. The water particles are now surrounding the sugar molecules which, because they are so small, cannot be seen in the water solution and they fit into the spaces between the water particles. Flour particles cannot be pulled away from each other by water and so they are insoluble in water. If you stir flour in water, the result is cloudy water. If you leave the mixture, eventually the particles of flour will drop out of the water and sink to the bottom of the container. They have not dissolved in the water. We say that the flour was **suspended** in the water. If you leave a mixture of sugar and water, the sugar will not sink to the bottom because it has dissolved.

Dissolving is fastest in hot water because the heat energy makes the particles move faster, thereby enabling them to mix more quickly and spread out more.

Soluble substances vary in how much can be dissolved in water (**saturation** level) because the forces of attachment in different substances vary.

PRACTICAL TASK PRACTICAL TASK **PRACTICAL TASK** PRACTICAL TASK

With children, try mixing a range of edible substances to water and sort them into 'soluble', 'insoluble' and 'suspended'. What difference does it make if you use hot water or if you stir the water? Expect descriptions of observations rather than explanation about the level of particles.

Changes of state

Melting

In a solid, the particles vibrate but are in fixed positions and held together by intermolecular forces. If more heat energy is given, then the vibrations can increase. When something solid melts, it means there is enough heat energy (and the amount will vary according to the type of material) for the movements to become so great that the forces between the particles are weakened and can no longer stick closely together.

The particles continue vibrating but, instead of being in fixed positions, they now roll over one another. The solid has melted to become a liquid. This is what happens when ice or butter melts. The temperature at which a solid melts is called the melting point of that material. If cooled sufficiently, the particles will slow down. They will not have enough energy to move around and so the forces of attraction are strengthened.

Evaporating

If heat energy continues to be given, the movement of the particles becomes so great that some of those on the surface fly off into the space above the liquid. Only those on the surface can fly off because they are attracted by other particles on one side only. (Particles beneath the surface are attracted by other particles on all sides.) The liquid has begun to change to a gas. As a gas, the particles are not held together and they move off in random directions, eventually filling all the space available and colliding with the sides of that space. If cooled sufficiently, the particles will slow down and begin to get closer together and start rolling over each other (condensation). If cooled further, the particles will slow down more and the forces between the particles will be strengthened.

RESEARCH SUMMARY RESEARCH SUMMARY RESEARCH SUMMARY **RESEARCH SUMMARY**

Many children will describe dissolving in terms of the disappearance of the solute, e.g. when sugar is added to water the sugar disappears or it evaporates. By Key Stage 3, the most common explanation for sugar dissolving in water is that the sugar has become liquefied, i.e. it has melted. Driver (1989) and many other researchers have found that children believe dissolving and melting are very similar or synonymous phenomena. Krnel *et al.* (1998) found that:

- children will use an atomistic explanation of dissolving, i.e. they will use words like 'break'. They will describe the sugar as breaking from larger particles into smaller particles;

- they might use the term 'push' – the water pushes apart the particles of sugar;

- children often confuse physical and chemical change in dissolving, e.g. some will claim that there is always a chemical reaction;

- they have problems with the conservation of weight. Because sugar, as they claim, disappears when it's added to water, the sugar and water separately will weigh more than when you add the sugar to the water and it dissolves. They tend not to have this misunderstanding with insoluble materials or when one liquid is added to another liquid.

Although some children are able to describe dissolving in terms of matter being broken down from large particles into smaller particles, most children have great difficulty explaining in terms of the atomic, ionic or molecular level.

Oversby (2000) suggests using coloured soluble materials because white sugar and salt become invisible when dissolved in water, reinforcing the misconception that they have disappeared.

Some substances, e.g. bromine, will go directly from solid to gas if their particles have sufficient energy to break away. This is known as **sublimation** and is not very common.

The three states of matter and the changes which take place between them are represented in Figure 7.1.

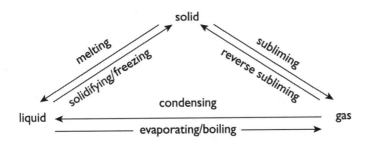

Figure 7.1 Changes of state

How do melting and evaporating relate to water?

The same rule applies. When there is sufficient heat energy the particles in ice will vibrate so much that they break away from their positions and roll around one another. The state is now called water. The melting point of ice is 0°C. If water is cooled (e.g. by placing in a freezer), then the particles will bond together in a rigid structure. The state is now called ice. The **freezing** point of water is 0°C.

Water is a rather odd material. Most materials expand when they change from a solid to liquid (the particles are a little further apart). Conversely, most materials contract when they change from a liquid to solid. That is, their mass does not change but their volume (and hence, their density) changes. Water does the oppo-site, i.e. it contracts when it melts and the temperature rises from 0 to 4°C and then it expands above 4°C. This means that water at 4°C takes up less space which means it is at its most dense. So, water at 4°C will sink in water which is colder or warmer. When water solidifies (freezes) it expands. This means that the same amount of water will occupy a bigger volume as a solid (ice) than as a liquid (water). This is why lettuce goes limp in a freezer. As the water in the cells of the lettuce leaf become solid, it expands and breaks the walls of the cells.

If the same amount is occupying a bigger space, this must mean that ice is less dense than water. Things will float on water if they are less dense than water. Therefore, ice will float on water. This is useful for living things in ponds and lakes because it means that the part of the water that changes to solid (ice) will always float on the surface of the water. In doing so, it will insulate the water beneath so that further cooling is slowed down. Thus, living things deeper in the water are protected. A disadvantage of the same property is that water which changes to ice in pipes will expand. If an opportunity to expand is blocked, then the force of the ice expanding will break the surrounding material (e.g. copper pipe) so that there is room for the expanded frozen water. When the ice melts, it contracts as it changes back to water leaving 'burst' pipes!

When heat is given to ice at 0°C it melts, but it doesn't get any hotter until all the ice has melted. All the heat energy is used to break apart the particles until they are all free to move around as a liquid. This is known as the latent heat of fusion: latent means hidden; fusion means melting.

RESEARCH SUMMARY RESEARCH SUMMARY RESEARCH SUMMARY **RESEARCH SUMMARY**

Krnel *et al.* (1998) summarised research into children's misconceptions, which found that:

- although many children could identify melting in water and understood the change from solid to liquid with water, they weren't able to generalise this into other substances that change from solid to liquid;
- melting is believed to always involve water, e.g. when melting paraffin wax, water is formed, when melting butter, water (or a similar substance) is formed;
- children frequently confused and mixed the words 'dissolving', 'melting' and 'turning into water', using them to mean the same thing.

PRACTICAL TASK PRACTICAL TASK **PRACTICAL TASK** PRACTICAL TASK

Try melting butter, water, chocolate and paraffin wax with a group of children. You could float small samples in dishes in a bowl of hot water. Compare their explanations or descriptions with Krnel's findings, above.

Evaporating and boiling

When water evaporates, it changes into a gaseous state: water vapour. This can happen at any temperature but it is faster and easier to see at higher temperatures. In a liquid, the particles are constantly moving around each other but some of the particles near the surface of the liquid will have enough energy to escape. If water continues to be heated, it will become so hot that there is evaporation inside the liquid and bubbles of water vapour form inside the water and rise to the surface (where they can escape into the space above). This is the boiling point and for water the boiling point is 100°C. If water is boiling and you continue to put in heat energy, this will not raise the temperature of the water. The energy will be used to turn more of the liquid into water vapour and the temperature stays at 100°C. This is known as the latent (i.e. hidden) heat of vaporisation.

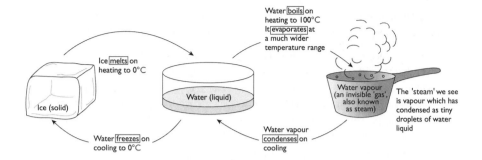

Figure 7.2 The water cycle

If you dissolve substances in water, this will lower the freezing point as well as raise the boiling point. Thus, putting salt on roads in winter means that any water (from snow or rain) on these roads will not freeze to become ice until much below 0°C. Similarly, salty water will boil above 100°C. Sea water (at atmospheric pressure) has a melting point of about –9°C and a boiling point of about 102°C.

How do we explain that mass is conserved in physical and chemical changes?

Chemical changes include such common occurrences as **burning** wood or mixing vinegar with sodium bicarbonate. New substances are made, but the total mass of the new substances is equal to the total mass of the substances involved in the chemical reaction (the reactants). Some children have difficulty with this when the reaction involves something that can't be seen, e.g. oxygen in air. So, for instance, when they see a **rusting** railing, they will think that there is less iron; the iron has been lost. Some children will find it difficult to understand that some of the iron has reacted chemically with the air to produce iron oxide:

iron + oxygen = iron oxide

RESEARCH SUMMARY RESEARCH SUMMARY RESEARCH SUMMARY **RESEARCH SUMMARY**

Krnel *et al.* (1998) and Johnson (1998) provided summaries of research about children's understanding of condensation and evaporation. At Key Stage 1, children say that the water disappears or goes into its immediate surroundings, e.g. when describing how a puddle evaporates, they will say that the water goes into the ground or it is absorbed by the flooring.

At Key Stage 2, children will tend to explain evaporation as something being transferred into the air or the sky, but it is somewhere away from themselves. Very few of them will describe water as changing into vapour that is dispersed into the air surrounding them. They might say the water has been displaced, e.g. it has been absorbed by the container or gone into the cooker. Even adults believe that water disappears during drying. Children tend to describe the gaseous state of water as invisible water particles, or a mixture of air and water, or water changed into air.

At Key Stages 3 and 4, children interchange water and air, so they might say that in boiling, bubbles are made of air (most common view), or they are made of heat, or that water has become air or that air has become water. When describing the visible water vapour from boiling, pupils tend to say that once it has lost its visibility it becomes or changes into air.

Moving on to the concept of condensation, the research shows that few children are able to accept that there is always water vapour in the air. Very few children at Key Stage 2 could accept that water vapour could change into liquid water. Children find it even harder to explain condensation of water on cold surfaces and many believe that condensation is a result of displacement, i.e. the water that was inside the vessel has moved to the outside and that is why there are water droplets on the outside of a cold container.

Analysis of Key Stage 2 SATs for 2005 suggested that pupils need the opportunity to 'recognise the role evaporation plays in a variety of everyday contexts' (QCA, 2005, p2).

The total mass of iron oxide will be equivalent to the iron plus the oxygen involved in the reaction. Many, but by no means all, chemical reactions result in irreversible changes, whereas physical changes are normally reversible. For example, burning wood results in chemical reactions with gases in the air and it is impossible to regain wood from ash plus the other new compounds made.

When new substances are formed as a result of a chemical reaction, those new substances can be completely different from the reacting substances. They could be of a different state, e.g. two liquids could react to form gases or solids. They could be less or more reactive and can react differently to the original reacting substances.

Usually, to show a chemical reaction, a chemical equation is written and the balance of reactants against products is shown mathematically at the smallest, atomic levels:

$$2Mg \quad + \quad O_2 \quad \rightarrow \quad 2MgO$$

| two atoms of magnesium | two atoms of oxygen | two molecules of magnesium oxide, each containing one magnesium and one oxygen atom |

RESEARCH SUMMARY RESEARCH SUMMARY RESEARCH SUMMARY **RESEARCH SUMMARY**

Since many children do not hold a scientific view of a substance, Johnson (2000) concludes 'they have no means of recognising chemical change: the possibility that substances might change into other substances is not something they can even begin to think of'. Johnson (2002) found that Year 7–Year 9 pupils were reluctant to accept that the products of a chemical reaction could be a new substance in its own right, although there was some acceptance of this idea after direct teaching.

$$HCl \quad + \quad NaOH \quad \rightarrow \quad NaCl \quad + \quad H_2O$$

| hydrochloric acid | sodium hydroxide | sodium chloride | water |

In summary, we can say that the total mass of the reacting substances (the reactants) is equivalent to the total mass of the new substances formed (the products).

The behaviour of metals

Earlier, some examples of chemical reactions were given, some involving metals. Often elements in the periodic table, like **metals**, will react chemically in broadly predictable ways with other substances. There are 88 elements in the periodic table that are classed as 'metals'.

Metals tend to:

- be solid at room temperature (mercury is an exception and is liquid);
- have high melting and boiling points;
- have loose electrons in the outermost energy level;
- donate electrons to become positively-charged ions;
- be good conductors of electricity and heat;
- be malleable: that is, they can be bent without snapping;
- have strong bonds, making the metal strong and tough;
- be grey in colour but, when cut or polished, are usually shiny.

A few are magnetic (iron, nickel and cobalt).

Metals react chemically in certain predictable ways. However, there is a hierarchy of reactivity of metals, from potassium as the most reactive, to gold as the least reactive.

REFLECTIVE TASK

Why is a piece of copper jewellery more likely to tarnish than a piece of gold jewellery?

In nature, the reactivity of a metal is an indication of how likely it is to be found uncombined with other substances. The more reactive a metal is, the more difficult it is to extract from its ore. The least reactive metals are likely to be found in rocks in their pure form e.g. gold and silver.

1. Metals will often react chemically with **oxygen** (e.g. from the air) to make **oxides**:

    | Metal + oxygen | = metal oxide |
    | Copper + oxygen | = copper oxide |

Metal oxides are **bases**, i.e. they have a pH greater than 7 and react with acids to form salts and water.

For more information on acids and bases, see below.

2. With water, metals usually react to form **hydroxides** and **hydrogen**:

    | Metal + water | = metal hydroxide + hydrogen |
    | Calcium + water | = calcium hydroxide + hydrogen |

3. When metals react with an **acid**, **hydrogen** is again produced, along with a **salt**:

    | Metal + acid | = salt + hydrogen |
    | Zinc + sulphuric acid | = zinc sulphate + hydrogen |

Some more predictable chemical reactions: the behaviour of acids and bases

An acid is a compound that contains hydrogen and usually dissolves in water, forming hydrogen ions. Well known acids include lemon juice (citric acid), vinegar (ethanoic or acetic acid), vitamin C (ascorbic acid), sulphuric acid, hydrochloric acid and nitric acid. Most acids have a sour taste, but don't use this as a universal test!

A base is a substance that neutralises an acid, producing salts and water. Bases are usually metal oxides, hydroxides and carbonates. A base that dissolves in water is known as an **alkali**. Well known alkalis include soap, washing powder, oven cleaner, ammonia, sodium bicarbonate and sodium hydroxide. Alkalis tend to feel slimy or soapy.

While some acids and alkalis can be safe to handle and taste, many are corrosive and can cause serious damage to skin.

The acidity or alkalinity of a substance can be measured using **pH**, which has a scale of 1 to 14, with 1 being the most acidic, 14 the most alkaline and 7 being neutral. Acids have a pH of less than 7 and alkalis have a pH of more than 7. Moving one point on the pH scale represents a ten-fold change in acidity or alkalinity. Thus, a substance with a pH of 3 is 1000 times more acidic than a substance with a pH of 6. The pH of pure water is 7, although water isn't the only common neutral substance. Other examples include olive oil, whisky and sugar.

pH	1	2	3	4	5	6	7	8	9	10	11	12	13	14
		strong acids			weak acids		neutral			weak alkalis			strong alkalis	

Acids can be diluted by adding water. Less water makes acids more concentrated. An acid is described as strong or weak, dependent on the amount of hydrogen ions produced when added to water (strong acids produce many more hydrogen ions). pH will measure the strength of an acid or alkali, not its concentration when diluted in water.

Common indicators used in school to detect acids and alkalis include litmus paper which turns red when touching an acid and blue with alkali. Universal indicator, which is a mixture of dyes, can be used to measure the pH of a substance (i.e. its strength) by matching the colour change to a chart.

PRACTICAL TASK PRACTICAL TASK **PRACTICAL TASK** PRACTICAL TASK

You could make your own indicator. The liquid from boiled red cabbage can be used as an indicator of acidity since it changes colour when added to an acid or alkali. Try adding some of it to the 'kitchen' materials mentioned above when it has cooled. Keep a record of the colour changes, e.g. by 'painting' them on white paper.

Acids can corrode, and thereby weaken, metals:

Acid + metal   =   salt + hydrogen

Acids react with an alkali or a base to form a salt and water:

Acid + metal oxide   =   salt + water
Acid + metal hydroxide   =   salt + water
Acid + metal carbonate   =   salt + carbon dioxide + water

The alkali or base neutralises the acid:

H^+(aqueous) + OH^-(aqueous) = H_2O (liquid)
from acid   from alkali   water

Acid   +   alkali = neutral substances

Acid produced by our stomach is neutralised by an alkaline substance, bile, when the stomach contents move to the small intestine. If excess acid is produced by the stomach, then the subsequent indigestion can be relieved by taking an indigestion remedy that helps to neutralise the acid and form a salt:

Hydrochloric acid   + magnesium oxide   = magnesium chloride   +   water
Excess stomach acid   base in the   the salt   water
                              indigestion remedy

Soils can be acid, alkaline or neutral and certain plants thrive in acid, alkaline or neutral soils. Keen gardeners can use a pH meter to test the acidity of soil and then alter the acidity by 'liming' the soil, or use the information to decide which plants are likely to grow best.

What about burning?

Burning is a type of chemical reaction. Burning depends on **fuels** such as coal, oil, wood or natural gas. When fuels burn, they combine with oxygen to make new chemicals (oxides). Chemical reactions like this give out heat to their surroundings. To start a fire, you need fuel, oxygen and usually heat. This is often shown as the 'triangle of fire' (see Figure 7.3).

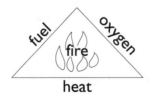

Figure 7.3 Triangle of fire

Fossil fuels are mostly made of carbon and hydrogen and the chemical reaction you get is:

fuel   +   oxygen   →   carbon dioxide   +   water   +   heat

Exothermic and endothermic reactions

Chemical reactions which give out heat to their surroundings are called **exothermic reactions**. Burning is not the only way to get exothermic reactions. If you add water to plaster of Paris, as it sets, its temperature rises. The chemical reaction between plaster of Paris and water is exothermic. This is because, when water is added to dry plaster of Paris powder, the bonds in the molecules of the plaster of Paris and the liquid water are broken to form the new hydrated plaster of Paris. The total **bond energy** needed between atoms in the new chemical is less than in the original plaster of Paris and water. The excess bond energy is transferred to the surroundings in the form of heat energy.

RESEARCH SUMMARY RESEARCH SUMMARY RESEARCH SUMMARY **RESEARCH SUMMARY**

Krnel *et al.* (1998), when summarising research, found many children believed that if, for instance, wood is burning and water appears on the wood, it is because the wood has been damp. Similarly, gases that have formed during combustion were believed to be already in the material that is burning, and smoke that is formed during combustion was in the wood before combustion.

Many children are unsure about the role played by oxygen because it isn't noticed. Being invisible, it isn't seen as being part of combustion, or is thought to enable combustion but not actively take part.

Often, children think that during combustion there are new products being made and these are completely different from the reactants, the materials that were burned. If there is a change, for instance a colour change or a change in form, they believe there has to be a change in mass as well. So, many children think a substance will change in weight, particularly when it changes from a solid into a liquid and further on into a gas, and that the total mass of the substances will be much less, especially the gases.

Johnson (2002) confirms that combustion is a very demanding concept, particularly in the context of the common activity of lighting a candle. Very few pupils understand the role of wax and oxygen in combustion and have little notion that chemical reactions are involved and new substances are formed. He warns that 'combustion must be regarded as one of the last things we should expect our pupils to understand' (p1053). However, direct teachings of particle theory to pupils in Year 7–Year 9 helped some to begin to understand what a gas is and the nature of chemical change and chemical reaction.

So, what happens when a candle is lit? (Adult-only explanation!) The candle wax is made up of carbon and hydrogen. Part of the candle melts when the wick is lit. The temperature will rise to a point (flashpoint) at which the carbon and hydrogen atoms split and react with oxygen in the air to form new chemicals, carbon dioxide and water. The liquid wax vaporises to form a gas which burns as a flame. White hot particles of carbon give the flame its brightness and the flame is composed partly of the incandescent particles of carbon and partly of unburnt gas. If you blow out the flame, the liquid wax solidifies (a reversible change). However, the vapour disperses into its surroundings; it does not condense to form a liquid (an irreversible change).

Some reactions take in heat energy from their surroundings and they are known as **endothermic reactions**. If you add vinegar to sodium bicarbonate, there will be lots of fizzing (as the gas, carbon dioxide, is being formed) and the mixture will drop in temperature; it will feel cold. This is because, when vinegar is added to sodium bicarbonate, the bonds in the molecules of vinegar and the sodium bicarbonate are broken to form the new chemicals, which include the gas carbon dioxide. The total bond energy needed between atoms in the new chemicals is more than in the original vinegar and sodium bicarbonate. The extra energy needed is transferred from the heat energy in the surroundings to the bond energies of the new chemicals.

PRACTICAL TASK PRACTICAL TASK **PRACTICAL TASK** PRACTICAL TASK

Add some white vinegar to two teaspoons of sodium bicarbonate in a sealed plastic bag. Note how it feels, looks, sounds and smells. Try varying the proportions. Alternatively, try using a narrow-necked container over which a balloon can be stretched. Put the vinegar in the container and the sodium bicarbonate in the balloon. Stretch the balloon over the neck of the container and then allow the contents of the balloon to pour into the container.

RESEARCH SUMMARY RESEARCH SUMMARY **RESEARCH SUMMARY** **RESEARCH SUMMARY**

Often, children will describe chemical reactions in terms of physical changes so, for instance, rusting is a change in the ageing of the iron rather than a chemical change. Children also tend to focus on the beginnings and the ends of a reaction and don't notice the dynamic changes during the process. This also applies to combustion. They might just look at the physical visible properties and focus on those, before and after reaction.

Children often confuse chemical reaction with mixing. So, a chemical reaction between two gases could be described as a mixing of those gases. They would claim that there may be some changes, for instance in physical properties such as colour, but the substances are still the same as before the chemical reaction (Krnel *et al.*, 1998).

Geological changes

The physical and chemical changes described thus far occur over a relatively short period of time. Some physical changes, such as geological changes, happen over millions of years.

RESEARCH SUMMARY RESEARCH SUMMARY **RESEARCH SUMMARY** **RESEARCH SUMMARY**

Trend (2000) found that primary trainee teachers clustered the geological past into three main categories: extremely ancient, less ancient and geologically recent, and they had difficulty sequencing key geological events. Children (10–11 years old) identified two main categories: extremely ancient and less ancient (Trend, 1998).

The Earth can be divided into three main layers, core, mantle and crust. The core, the central part of the Earth, is at a depth below 2900km and is thought to be largely nickel and iron. The next layer, the mantle, is wrapped around the core and ranges in depth from 90–2900km. The Earth's crust is made of rock and lies on top of the mantle. The crust is relatively thin, varying from 6–11km under the oceans to 25–90km under the continents. Continental crust is largely granitic whereas oceanic crust is largely basaltic. Although there are places where the crust is exposed, such as cliffs and mountains, much of it is hidden beneath water or soil.

Rock is made of **minerals** which are naturally occurring chemical compounds. For example, iron pyrites is a mineral that is a compound of iron and sulphur. Minerals are originally formed from hot magma. They form regular lattice structures, i.e. crystals, the shape of the crystal being determined by the arrangement of the atoms. The vast majority of minerals are made of the following eight elements: aluminium, calcium, iron, magnesium, oxygen, potassium, silicon and sodium. Rocks can be made up of one or several minerals, for example granite is made up of the minerals quartz, feldspar and mica.

RESEARCH SUMMARY RESEARCH SUMMARY **RESEARCH SUMMARY** **RESEARCH SUMMARY**

When describing rocks, many children use an everyday definition of 'a boulder-sized fragment of hard, dense and dark coloured rock'. 'Crystal', 'stone' and 'pebble' loosely distinguish size and shape. 'Mineral' was hardly ever used in the context of rock (Happs, 1985). Rocks are large, dull and rough, whereas stones are small, smooth and round. Children have little understanding of the terms sedimentary, igneous and metamorphic in the context of how rocks are formed. Most children believed that land is mostly soil with the occasional rock. They lack an understanding of land having a thin covering of soil derived from the local, parent rock and the Earth being largely composed of rock (Russell *et al.*, 1993).

There are three main types of rock formation: **igneous**, **sedimentary** and **metamorphic**.

Igneous rocks are formed from molten rock (magma) that is forced up to the surface of the Earth's crust, for example through volcanoes or from hot magma forced up through the ground. Over time the rock cools and hardens. It contains minerals that form into crystals. Igneous rock is often hard and never contains fossils or forms layers. Granite, basalt and pumice are examples of igneous rock.

Sedimentary rock is formed when other rocks are weathered and eroded and the bits are carried by wind and water and are deposited as layers of sediment that develop over millions of years at the bottom of lakes and seas. The layers are cemented together by crystals and harden. Fossils can be found in some sedimentary rock, formed from the remains of animals and plants. One way of dating layers of sedimentary rock is by the presence of particular types of fossilised animals and plants. Chalk and sandstone are examples of sedimentary rock. Chalk is the slow build-up of tiny skeletons of sea animals, laid down about 100 million years ago. It takes about 1000 years to lay down a 1cm depth of chalk.

Metamorphic rock is igneous and sedimentary rock that has been changed by heat and pressure over a very long time. New minerals are formed. Some metamorphic rocks have layers or bands and the rocks are usually hard. Marble and slate are examples of metamorphic rock. Marble is a metamorphic limestone (sedimentary rock) and slate is a metamorphic mudstone (sedimentary rock) that has been heated and squashed.

A few examples of the ages of different types of rocks:

Name of rock	Type of rock	Possible location	Approximate age in years
Basalt	Igneous	Skye, Scotland	60,000,000
Granite	Igneous	Dartmoor, Devon	280,000,000
Limestone	Sedimentary	Derbyshire	320,000,000
Sandstone	Sedimentary	Leeds	300,000,000
Schist	Metamorphic	Scotland	450,000,000

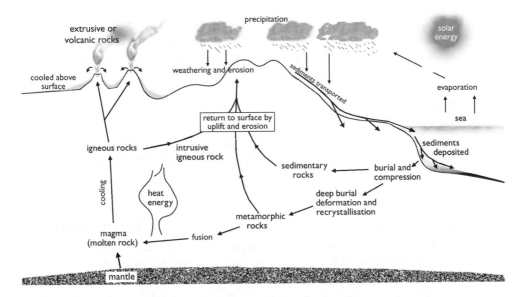

Figure 7.4 A simplified diagram of the rock cycle

The rock cycle describes the changing of the three types of rock from one form to another which occurs over many millions of years. Igneous rocks are eroded to form sediments, sediments are metamorphosed and may eventually be melted to become a new cycle of igneous rocks. Such changes are the results of a number of factors.

- Physical weathering – rocks can be broken down into smaller bits by various forms of physical weathering. Rock can break up because water in the rock freezes, thereby expanding and cracking it. As the cracks become bigger, bits of rock will break off. The surface of a rock can expand and contract as a result of temperature changes by day and by night. As a result, bits of rock will peel away from the main rock. Gravity can cause landslides.
- Chemical weathering – minerals are dissolved by chemical reaction with water as well as oxygen and carbon dioxide in the air. Acid rain is an example of chemical weathering. Rocks containing carbonates, like limestone, will react chemically with acid rain and eventually break down.
- Biological weathering – the result of disturbance and weakening of rocks by animals burrowing and plant roots (such as from trees) growing through rocks, forcing open cracks. Weakened rocks are relatively easy to break up.

- Erosion – worn down bits of rock are transported around the Earth by wind, water and gravity.
- Movement – movements in the Earth's crust can cause rocks to bend, buckle and scrape past each other.
- Deposition – layers of sediment are built up.
- Compression – layers are squeezed and compressed so that eventually sedimentary rock is formed.
- Heat/pressure – more squashing and heating turns rocks into metamorphic rocks.
- Melting – intense heat can melt rock, changing it to igneous rock.
- Cooling – molten rock can solidify.

There are three main divisions making up soil: bedrock, subsoil and topsoil. Parts of the bedrock break off and begin to break down into soil. Above the bedrock, the subsoil is composed of the smaller, broken pieces of bedrock. The roots of some plants can reach the subsoil and there will be the remains of some animals and plants at this layer. The topsoil is a mixture of weathered rocks, the remains of dead plants and animals, air, water and small living animals.

REFLECTIVE TASK

Ask a group of children to explain what soil is made from and note their responses. Show them a fossil and ask them to explain how it got into the rock. Note their responses. Compare these with the research summaries.

RESEARCH SUMMARY RESEARCH SUMMARY RESEARCH SUMMARY **RESEARCH SUMMARY**

Geological change was rarely mentioned by children (Russell *et al.*, 1993) but, when it was, the focus was on weathering by the Sun and by water. There was little awareness of large-scale geological change.

Children believe that soil is unchanging and has been there since the beginning of time. They might be aware of soil being part of a cycle involving rocks and clay but they are unclear how this might be (Happs, 1984). Similarly, mountains are large hills that have been around since time began. Some older children linked the formation of mountains to volcanoes, folding and faulting of the Earth's surface or to plate tectonics (Happs, 1982 in Russell, *et al.*, 1993).

A SUMMARY OF **KEY POINTS**

> **Dissolving can be explained at the level of particles.**

> **Most materials exist in a solid, liquid or gaseous state and can change physically from one state to another.**

> **The temperature at which each change of state can happen will vary considerably, depending on the material.**

> **Melting, evaporating, condensing, freezing, boiling and sublimating are all terms used to describe a physical change in state from one form to another.**

> **In chemical changes, the total mass of the new substances is the same as the total mass of the original substances involved in the chemical reaction.**

> Some elements and compounds can be grouped according to predictable chemical reactions.

> In physical changes, the mass of the material remains the same whether in liquid, solid or gaseous form; however, the volume that the same mass occupies can change and, therefore, so will its density.

> In some chemical reactions, energy, in the form of heat, is given out or taken in from the surroundings – this is a result of a change in the total bond energies required for making the new substances.

> Some changes take place over millions of years.

M-LEVEL EXTENSION > > > > M-LEVEL EXTENSION > > > >

Look again at the concepts dealt with in this chapter. Some, such as how everyday materials change when they are heated or cooled, are relatively simple to explore in the classroom, but others, including the longer-term geological changes in the rock cycle, are not as easy. Consider how the use of ICT may be able to help you to demonstrate some of these key aspects, for example through video clips and simulations. You may find it helpful to discuss this with colleagues including the subject leader for ICT.

REFERENCES REFERENCES REFERENCES REFERENCES REFERENCES

Driver, R. (1989) Beyond appearances: the conservation of matter under physical and chemical transformations, in R. Driver, E. Guesne and A. Tiberghien (eds) *Children's Ideas in Science*. Buckingham: Open University Press.

Happs, J. C. (1984) Soil genesis and development: views held by New Zealand students. *Journal of Geography*, 83(4), 177–80.

Happs, J. C. (1985) Regression on learning outcomes: some examples from the earth sciences. *European Journal of Science Education*, 7(4), 431–43.

Hesse, J. J. and Andersson, C. W. (1992) Students' conceptions of chemical change. *Journal of Research in Science Teaching*, 29(3), 277–99.

Johnson, P. (1998) Children's understanding of changes of state involving the gas state. Part 1: boiling water and the particle theory. *International Journal of Science Education*, 20(5), 567–83.

Johnson, P. (1998) Children's understanding of changes of state involving the gas state. Part 2: evaporation and condensation below boiling point. *International Journal of Science Education*, 20(5), 695–709.

Johnson, P. (2000) Children's understanding of substances, part 1: recognising chemical change. *International Journal of Science Education*, 22(7), 719–37.

Johnson, P. (2002) Children's understanding of substances, part 2: explaining chemical change. *International Journal of Science Education*, 24(10), 1037–54.

Krnel, D., Watson, R. and Glazar, S. A. (1998) Survey of research related to the development of the concept of matter. *International Journal of Science Education*, 20(3), 257–89.

Oversby, J. (2000) Good explanations for dissolving. *Primary Science Review*. 63, 16–19.

QCA (2005) *Implications for teaching and learning from 2005 national curriculum test*. Available at www.qca.org.uk/ks2app/ (accessed 3 Dec. 2008).

Russell, T., Bell, D., Longden, K. and McGuigan, L. (1993) *SPACE – Rocks, Soils and Weather*. Liverpool: Liverpool University Press.

Trend, R. (1998) Investigation into understanding of geological time among 10- and 11-year-old children. *International Journal of Science Education*, 20(8), 973–88.

Trend, R. (2000) Conceptions of geological time among primary trainee teachers, with reference to their engagement with geoscience, history and science. *International Journal of Science Education*, 22(5), 539–55.

FURTHER READING FURTHER READING **FURTHER READING** FURTHER READING

Archer, D. (1991) *What's Your Reaction?* London: The Royal Society of Chemistry.

Bell, D. and George, N. (eds) (1997) *Nuffield Primary Science – Understanding Science Ideas: a Guide for Primary Teachers*. London: Collins Educational.

Bethell, A., Dexter, J. and Griffiths, M. (1996) *Heinemann Co-ordinated Science: Chemistry*. Oxford: Heinemann Educational & Professional Publishing Ltd. Any other good textbook for GCSE science should enable you to extend your knowledge.

de Boo, M. (2000) *Laying the foundations in the early years*. Hatfield: Association for Science Education.

DfE (2011) *Teachers' Standards*. Available at www.education.gov.uk/publications.

Nuffield Primary Science (1995) *Materials – Teacher's Guide*, 2nd edn. London: Collins Educational.

Pendlington, S., Palacio, D. and Summers, M. (1993) *Understanding Materials and Why They Change.* Oxford: Oxford University Department of Educational Studies and Westminster College.

Taber, K. (2002) *Chemical misconceptions – prevention, diagnosis and cure. Volume 1: theoretical background*. London: Royal Society of Chemistry.

8
Electricity and magnetism

Curriculum context

National Curriculum programmes of study

At Key Stage 1, children should be taught about everyday electrical appliances, about simple series circuits, how a switch can break a circuit, to sort objects into magnetic and non-magnetic, and to recognise pulls and pushes as forces.

At Key Stage 2, children should be taught that some materials are better electrical conductors than others, to construct a range of circuits, that changing the number or type of components in a series circuit can make bulbs dimmer or brighter, how to use conventional symbols to represent electrical components, to compare materials on the basis of their magnetic behaviour, about the forces of attraction and repulsion between magnets, about the forces of attraction between magnets and magnetic materials, and how to measure forces and identify the directions in which they act.

Early Years Foundation Stage

Children must be supported in developing the knowledge, skills and understanding that help them to make sense of the world. Their learning must be supported through offering opportunities for them to undertake practical 'experiments'; and work with a range of materials. By the end of the EYFS, children should:

- investigate objects and materials by using all of their senses as appropriate;
- ask questions about why things happen and how things work;
- build and construct with a wide range of objects, selecting appropriate resources where necessary;
- find out about and identify the uses of everyday technology.

Introduction

Electricity cannot be seen but its effects are everywhere. Electrical flow produces heat, light, sound and movement. Modern life would be unthinkable without electricity, but the first commercial electric lamps started glowing only in 1880. Just in terms of the production of light, electricity is used by a wonderful range of devices, including light bulbs, fluorescent tubes, light emitting diodes, neon tubes, lasers and cathode ray tubes.

Electrons carry electrical charge

For more information on electrons, see the section 'What are materials made of?' in Chapter 6.

All materials are made up from tiny particles called **atoms**. In turn, atoms are made from even smaller particles including a nucleus and **electrons**. The nucleus of an atom has a positive electrical charge but electrons have a negative charge. These two opposites cancel each other out, leaving the atom as a whole electrically neutral.

This information is important when trying to understand electrical flow. **Electricity** can be thought of as a flow of electrons. The chemicals in a **battery** react together to produce a surplus of electrons at the negative terminal and a deficit of electrons at the positive terminal. Since electrons are negatively charged they are attracted to the positive terminal of the battery.

Many children and adults find the idea that materials are made up from particles difficult to accept. It contradicts our everyday experience that solid objects are made of continuous material and not tiny specks of material. The electrons, which circle the atom's nucleus like a cloud, have virtually no weight. The centre, or nucleus, of the atom is heavy in comparison with the electrons.

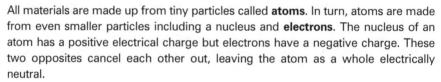

RESEARCH SUMMARY RESEARCH SUMMARY RESEARCH SUMMARY **RESEARCH SUMMARY**

Kibble (2002) reported a project involving primary children and trainee teachers and what they thought happened inside wires. The children and trainees were asked to draw what they would see if they were very small and inside the wire. The trainee teachers' ideas were categorised into four broad groups: in the commonest model the trainee shows a wave or sparks or both; a few had a moving particles model with charged particles moving through the wire; some trainees had no model at all. The children (aged 9–11) had similar conceptions to those held by the trainee teachers. They frequently drew waves and sparks. The children also had a distinctive model, which Kibble called the 'energetic model'. This had lively words such as 'light', 'hot', 'noise', 'power' and 'zap'. Kibble proposed activities designed to teach the standard scientific moving electron model.

Teachers have become increasingly aware of the range of analogies, e.g. Asoko and de Boo (2001). Mulhall *et al.* (2001) pointed out that the abstract concepts of electricity are both difficult to understand and particularly dependent on models, analogies and metaphors.

Asoko (2002) analysed the problems pupils have with the concept of electrical flow. Firstly, if the battery supplies electricity to the bulb, why do we need two wires? Secondly, if the electricity goes round the circuit and comes back to the battery, why does it go flat? These ideas are discussed later in this chapter.

Good and bad conductors

A **conductor** is a material which allows **electricity** to flow through it. Since electricity is the flow of electrons, this means that conductors must have electrons which are free to move within the material. All metals have these free electrons, so they all conduct electricity. Non-metal materials such as carbon, silicon and germanium are semi-conductors (or moderate conductors). In electrical **insulators** none of the electrons is free to move between atoms. This means that electrical flow is not possible (see Table 8.1).

Good conductors	Moderate conductors	Poor conductors	Insulators
all metals, e.g.: copper aluminium zinc magnesium gold silver nickel	silicon carbon germanium	water skin	rubber wood plastic pottery dry air

Table 8.1 The conductivity of different materials

Many children and adults think that the insulation covering wires helps electricity flow along the wire. This is not the case. The covering simply stops bare wires touching each other and leading to a **short circuit**. This idea can be checked by comparing two **circuits**, one of which incorporates a length of insulated wire and the other which uses the same length and type of wire with the insulation removed. There will be no difference in the brightness of the two bulbs.

It is easy to make mistakes about which items conduct electricity and which are insulators, because so many metals are coated in clear varnish. Metal coat-hangers, for instance, are coated to prevent them marking clothes. A vigorous rub with sandpaper removes the varnish, baring the metal underneath. The bright copper windings in a motor or a bell seem to be bare wire. This is not the case – the wire is coated with varnish to prevent short circuits.

Electrical resistance

Electrical **resistance** is the measure of the difficulty that electrons have in flowing through a material or object. This difficulty is caused when moving electrons (the electrical current) bump into the atoms which make up the conductor. The more atoms which are struck in this way, the greater the resistance of the material. When an atom is bumped into by an electron, it vibrates and it is this vibration which represents the raised temperature of the conductor.

In summary, the amount of resistance depends on three factors.

1. The material the electricity is flowing through. Copper, silver and gold are the best conductors, while lead and iron are less good. This explains why wires are made from copper (the cheapest and best conductor). Good conductors have low resistance. Poor conductors have high resistance. All insulators have complete resistance.
2. The thickness of the material. A thick wire will let electricity flow with far less resistance than a thin wire. This explains why thin wire flexes are good enough for table lamps, which need only a small flow, but thicker flexes are needed for kettles and heaters, where the flow of electricity through the heating element is much greater.
3. The length of the resistor. A piece of wire double the length of a similar wire has double the resistance. This can be seen when comparing the brightness of a bulb in two simple circuits; one using 50cm and the other 5m of wire.

There are perfect conductors of electricity which offer no resistance to the flow of current. Unfortunately, these exotic materials work in laboratories only at very low temperatures (lower than –200°C). When these devices can be made to work at room temperature, the way we use electricity will change. Until then, in the everyday world, whenever electricity passes through a conductor it encounters resistance.

The wires which supply electricity to our houses and appliances are made of copper. We do not want these copper wires to become warm, so they are thick. Where a large flow of electricity is allowed by an appliance such as a heater, thick wires reduce resistance. However, even thick short copper wires will resist electricity to some extent and become a little warmer as electricity flows through them. Where there is high resistance to the flow of electricity, much of the electrical **energy** is converted into heat. This is precisely what is required in a light bulb. The **filament** of a traditional bulb is manufactured to maximise the heating effect of an electric current. A filament is:

- very thin (thinner than a human hair);
- very long (often coiled to maximise length);
- made of tungsten (a metal with a very high melting point and not one of the best conductors).

REFLECTIVE TASK
BEFFECIIAE IV2K

There are many analogies that illustrate the idea of electrical resistance. One likens it to a traffic jam where many cars (electrons) are trying to go up a narrow street. Another relies on the contrast between running along the road and running in mud. Which one helps you picture resistance? See Asoko and de Boo (2001).

Resistance is measured in ohms (Ω). To find the resistance of a light bulb two readings have to be taken.

- The electric current flowing in the circuit needs to be measured using an ammeter. Current is the same throughout the circuit. Current is measured in amps.
- The voltage across the light bulb is measured using a voltmeter. Voltage is measured in volts.

Once the two readings have been taken it is possible to calculate the resistance of the bulb. This is done by dividing the **voltage** by the current:

Resistance = $\dfrac{\text{Volts}}{\text{Amps}}$

In the circuit shown in Figure 8.1 the resistance of the bulb is:

$$\frac{1.5\text{v}}{0.2\text{A}} = 7.5\ \Omega$$

In a simple circuit the wires hardly resist at all compared with the bulb. This is because the wires are very thick compared with the filament of the bulb. In effect all the resistance of the simple circuit occurs in the filament of the bulb. It is useful to remember that the filament resists the flow of electricity. This helps to explain why two bulbs in **series** are dimmer than a single bulb; if you put two light bulbs in series you are doubling the length of the resistor, so doubling the resistance. This reduces the flow of electricity in the circuit.

Variable resistors are made from materials which are moderate conductors of electricity. They are sometimes found as dimmer switches on lights, volume controls on radios and speed controls on power drills and toy cars. Many variable resistors use carbon or windings of low conductivity metals.

PRACTICAL TASK PRACTICAL TASK **PRACTICAL TASK** PRACTICAL TASK

Set up a simple circuit with a bulb and a battery. Test different lengths of wire to see if the length affects the conductivity of the wire. Ordinary wire will not have any effect.

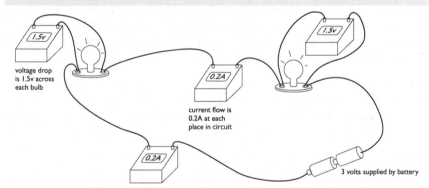

Figure 8.1 Circuit with two bulbs, two voltmeters and two ammeters

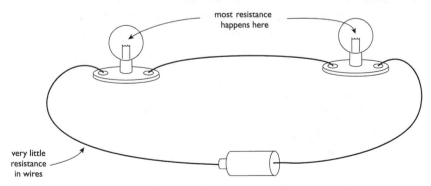

Figure 8.2 Two bulbs in series are dimmer than a single bulb

Electrical current which is forced to travel through a long piece of carbon encounters high resistance. Pencil leads are made from graphite, which is a form of carbon. If electricity in a circuit with a single battery and bulb has to pass through a propelling pencil lead, the graphite will resist the flow of electricity. The greater resistance cuts down the flow of electricity causing the bulb to glow less brightly. The brightness of the bulb can be varied by changing the length of graphite in the circuit. This is exactly the same principle on which mains lamp dimmer switches work.

As a resistor heats up, its resistance increases. This means that a light bulb which is glowing brightly will have a high resistance. By placing two bulbs in series the current is reduced resulting in lower temperatures in both bulbs. A dimly glowing bulb has a lower resistance than a bright bulb.

There are many types of special resistors. A light dependent resistor (LDR) has a high resistance when no light is falling on it and a low resistance when it is light. If an LDR is in a circuit with a buzzer, an alarm will sound if, for instance, a burglar's torchlight or car headlights fall on it. A thermistor is a resistor which decreases in its resistance as it gets hot. An alarm can be made to sound if the device is in a fire.

Electrical current

The flow of electrons produces electrical current. Current is measured in **amps** (A). A large mains current might have a current of 10 amps. Typically, in classroom experiments, batteries provide a flow of approximately 0.2A. Even this tiny flow involves the movement of many millions of electrons around the circuit. Electrons cannot leave the circuit nor can they bunch up or space themselves out. The flow of electrons everywhere in the circuit is instantaneous once a battery is connected. The current is equal in all parts of the circuit since there is the same rate of flow of electrons everywhere in the circuit. Electrons are not used up – it is the energy of the flow which is converted into heat or light.

If a circuit splits into two, and one of the paths is easier for electricity to flow along, more of the electricity will follow the easier path. Put in terms of resistance, electricity will always flow down the path of least resistance. Where a path offers virtually no resistance to the flow of electricity, this is referred to as a short circuit.

For more information on resistance, see the section on electrical resistance earlier in this chapter.

If there are two paths for electrons to flow down, they are referred to as being parallel. Even though the National Curriculum programmes of study do not explicitly refer to **parallel circuits**, it is likely that some children will already know about them or make a parallel circuit by accident. Notice that in Figure 8.3 there is 0.3 amps of flow down one path (c) and 0.2 amps flow down the other parallel path (b). The flow in the common wires (a) must be 0.5 amps to supply them both.

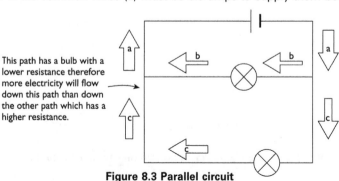

This path has a bulb with a lower resistance therefore more electricity will flow down this path than down the other path which has a higher resistance.

Figure 8.3 Parallel circuit

RESEARCH SUMMARY RESEARCH SUMMARY RESEARCH SUMMARY **RESEARCH SUMMARY**

Classic research by Osborne and Freyberg (1985), repeated by Chang *et al.* (2007), asked children what they thought happened to electrical flow. They were presented with four possibilities.

1. There is more current in the wire supplying the bulb and less in the return part of the circuit.

2. There is the same flow in supply and return wire.

3. There is the same flow from both ends of the battery and these two flows clash in the bulb.

4. There is a flow only in the supply wire and the return wire is unnecessary.

Most children in the primary phase believed option 1 to be the most likely. Only approximately 20 per cent felt the correct explanation (option 2) was believable. Clearly, most children felt that current was used up in the bulb. That is quite understandable since common sense might suggest this. Indeed, the scientific idea that the energy of the flow is changed into heat and light has many similarities with the children's ideas about current.

Series and parallel circuits

There are important differences between a circuit with two bulbs in parallel and a circuit with two bulbs in series.

In a series circuit

● There is only one path through all of the components in the circuit.

● When one bulb is unscrewed the other one goes out. This is a result of the electricity not having a complete circuit since it cannot now flow through the filament of the unscrewed bulb.

● The battery will last longer as the current (the flow of electricity) is considerably reduced by the increased resistance of the two bulbs in series.

● If the two bulbs have identical resistance they will glow with the same brightness.

● If one of the bulbs has a higher resistance than the other, the higher resistance bulb will glow more brightly. This is surprising! It happens because the bulb with higher resistance converts more of the energy of the flow into heat and light.

> NOTE: The brightness of the bulbs does not depend on the position. If the positions of the two bulbs which glow unequally are swapped the same bulb still glows more brightly.

In a parallel circuit

● There are several paths that the electricity can take.

● Either bulb can be unscrewed without affecting the other. This is because the electricity can use the other path in the circuit.

● The battery will be able to supply electricity for a relatively short time as there are two pathways for electricity to flow down.

● If one of the bulbs has a higher resistance than the other, the higher resistance bulb will glow less brightly.

> NOTE: The higher resistance bulb glows less brightly than a lower resistance bulb in a parallel path. This is because the pathway with the lower resistance bulb allows more electricity to flow down it than the parallel path.

For more information on resistance, see the section on electrical resistance earlier in this chapter.

The information printed on the screw of the bulb usually gives the current and working voltage, so you can work out the resistance of a bulb as shown below:

Bulb (a) 6 volts/0.2 amps...this bulb has a resistance of $\dfrac{6}{0.2} = 30\,\Omega$

Bulb (b) 2.5 volts/0.2 amps...this bulb has a resistance of $\dfrac{2.5}{0.2} = 12.5\,\Omega$

Bulb (c) 2.5 volts/0.1 amps...this bulb has a resistance of $\dfrac{2.5}{0.1} = 25\,\Omega$

PRACTICAL TASK PRACTICAL TASK PRACTICAL TASK PRACTICAL TASK

Try this useful activity as an analogy to help children to understand electrical flow. Place a large table or block of desks pushed together in the middle of the room. Ask the children to move slowly at random around the table. Explain that this is similar to how electrons in a conductor might move between atoms. Stand near to the children, attracting the 'electrons' with your right hand (positive) and repelling them with your left hand (negative). The previously randomly moving electrons now move in only one direction. The harder the battery works, the faster the electrons move around the table (circuit).

Now place a chair close to the table and direct the children to squeeze between the table and the chair – this gap mimics the resistance of a bulb filament. Draw the children's attention to the fact that the flow of electricity in the whole circuit dramatically declines. However, when you add a second pathway for the electrons to go down, the flow increases again. This shows the children the effects of a parallel circuit.

Voltage

Voltage is measured in volts (V). It is the difference in the potential energy between the positive terminal and the negative terminal. Analogies with water are useful. Imagine a flow of water just before it reaches a steep drop. There is a difference in the energy between the water at the top and the water in the pool at the foot of the falls. The water can be made to do work as it falls, such as turn a waterwheel.

For more information on potential, see the section on chemical changes and electricity in Chapter 9.

The higher the voltage the greater the amount of work the current can do – in other words, the hotter the bulb filament will become or the faster the motor will turn. A simple cell (often referred to as a battery) has a potential difference of 1.5V between the positive and negative terminals. When two or more cells are put in series the voltage of the batteries is added. Higher voltage results in more power.

$$1.5V + 1.5V = 3V \qquad 1.5V + 1.5V + 1.5V + 1.5V = 6V$$

To measure voltage a voltmeter is used. This device measures the potential energy of the electrons at different points in the circuit. The difference between the two ends (terminals) of a torch cell is 1.5V. This difference comes from the stored chemical energy of the cell. The cell does work by raising the charge of the electrons to a higher potential or voltage.

Please note that many of the connections in simple circuits are rather poor. Crocodile clips and the screws in bulb holders rarely make snug contacts. This results in some of the voltage being used to get across the wobbly contacts. So, experimenters should not be surprised if the sum of the voltage drop across all the conductors does not equal the voltage of the cell or battery.

Matching bulbs and batteries
Matching the voltage of the bulb to the battery is very important. If a high voltage bulb, e.g. 6V, is used with a single cell it will hardly glow at all. However, a 2.5V bulb connected to a 1.5V cell or a 3V battery will glow satisfactorily. Problems arise when a low voltage bulb has a high voltage passed through it. For instance, a 6V battery will melt the filament of a 2.5V bulb. This melting is commonly referred to as blowing the bulb. Problems like this can be avoided by checking the voltage on the screw of the bulb.

Cells and batteries

A single round battery is more correctly called a **cell**. Batteries are formed when a number of cells are grouped together, rather like hens in battery cages or a battery of guns. The simplest cell consists of a piece of zinc and a piece of copper put into an acid solution such as vinegar. This arrangement will not provide enough current to make a bulb glow, but two of them in series may light an LED.

Once all the chemicals in a non-rechargeable cell have reacted together, no more electricity can be produced and the battery is worn out. Rechargeable batteries rely on a flow of electricity to reverse the chemical reactions. This puts the chemicals back to their original state, allowing them to react again causing a further electrical flow.

PRACTICAL TASK PRACTICAL TASK **PRACTICAL TASK** PRACTICAL TASK

Key Stage 1 children can experiment with two cells placed in series to see how they increase the voltage if connected correctly. They will cancel each other out if connected incorrectly.

For demonstration by the teacher only: show how very brightly a bulb will glow if you connect four or five cells in series. Eventually the filament will melt, breaking the circuit.

Power

Power is the rate at which energy is transferred. It is measured in **watts** (W). Electrical power is calculated by multiplying the flow of electricity (amps) by volts.

$$\text{watts} = \text{amps} \times \text{volts}$$

All mains devices such as bulbs, heaters and TVs are rated showing their power in watts. A higher rating shows the device can convert energy more quickly than a device with a lower rating. A vacuum cleaner which is rated at 2000W is 33 per cent more powerful than a cleaner rated at 1500W. It is likely to have approximately 33 per cent stronger suction. A 2000W (2kW) fire will convert electricity into heat twice as quickly as one rated at 1000W (1kW).

Electrical apparatus	Power (W)	Current (A)	Voltage (V)
Torch bulb	0.3W	0.2A	1.5V (single cell)
Mains bulb	60W	0.25A	240V (mains voltage)
Heater	3kW (3000W)	12.5A	240V
Electric shower	8kW	33A	240V

Table 8.2 Power ratings of various electrical devices

Table 8.2 shows why separate circuits are needed for lighting, heating and shower circuits. Devices such as torch bulbs convert power at a very slow rate. Large heating devices convert electrical power at a very high rate as shown by their rating. The wires needed for a lighting circuit are usually thin since they are carrying only a relatively modest flow of electricity.

Electricity companies sell electricity by units. These units are equivalent to one kilowatt used for an hour. A device such as a 2000 watt (2kW) fire will use 6kW hours in three hours. A kettle which can convert electricity into heat at the rate of 3000W (3kW) will use 9kW hours of electricity in three hours.

Work

For more information on work, see the section on electrical energy in Chapter 9.

Electricity can do work. Electric motors can move things and heaters can increase the temperature of objects. Work is measured in joules (J). One joule of work is done when one watt of energy is used for one second.

$$\text{Energy used by a 3kW kettle in 10 minutes} = \text{power (watts)} \times \text{time (seconds)}$$
$$= 3000 \times 600 \text{ secs}$$
$$= 1\,800\,000 \text{ J}$$

Electrical energy changes

The wires supplying a table lamp or a hairdryer are quite thick. They are also made of an excellent conductor such as copper. This allows electricity to pass through them easily (with little resistance). However, a hairdryer's heating wires are made to resist electricity and to heat up. In the case of the heating wires, the metal used is often an alloy of nickel and chrome which can resist high temperatures and is not as highly conductive as other metals. As electrons struggle through the wire they bump into the atoms of the wire causing the element to become hot and glow.

Components and circuit symbols

Figure 8.4 shows the standard set of symbols for electrical devices.

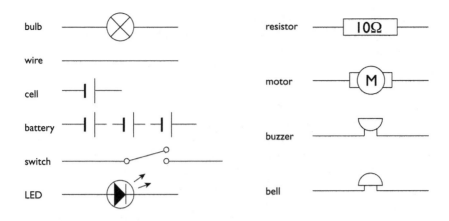

Figure 8.4 Symbols for electrical devices

Magnetism

Magnetic poles

Every magnet, no matter how small, has a north and a south pole. If a magnet is broken, it forms two separate magnets, each with a north and a south pole of their own. This observation is explained by the theory that all magnetic materials are divided up into magnetic domains. A magnetic domain is an area of the material which acts like a miniature magnet. In an unmagnetised piece of steel, these domains point in random directions. Once magnetised (see below for the methods), the domains line up, thus combining the magnetic effect of all the domains. New magnets can be created in two main ways.

1. Steel or iron objects, such as needles or nails, can be touched or stroked with an existing magnet (see below for compass needles).

2. Steel or iron objects can be placed in a solenoid (windings of wire through which a strong electric current flows).

In the case of steel, this magnetism lasts for a long period resulting in a permanent magnet. However, in the case of iron, the magnetism disappears immediately it is no longer in contact with the existing magnet or the electric current in the solenoid is switched off. Steel is ideal for compass needles and iron is ideal for making electromagnets which can be turned on and off quickly.

To remove the magnetism of a steel needle, it can be dropped on a table. This jumbles up the magnetic domains, returning them to a randomly aligned state. High temperatures will also cause a permanent magnet to lose its magnetism.

The pole of a magnet is the place where magnetism is strongest. North poles attract south poles but repel other north poles. This is summed up as the rule 'like poles repel, unlike poles attract'. On some magnets the poles are on the sides rather than the ends. Round magnets with holes in the middle (polo magnets) have their poles on the faces allowing enjoyable games with several polo magnets hovering above each other on a wooden rod. Normally bar magnets have their north-seeking poles painted red.

Compasses

The principle of the magnetic compass was discovered by the Chinese. Using a naturally occurring ore called lodestone, the Chinese, and later Viking mariners, were able to find north even when the sun or stars were obscured. Tudor sailors learnt to make a compass needle by stroking a steel needle with a piece of lode-stone. The magnetised needle was then floated on a cork and pointed north. The north magnetic pole of a bar magnet, the north-seeking pole, points north. It acts like this because the Earth behaves as if it has a bar magnet along its axis. Since north-seeking (north) poles of magnets are attracted to magnetic south poles, then the Arctic must act like a south magnetic pole and the Antarctic a north magnetic pole. This is, at first reading, a little puzzling and needs to be completely clear in the teacher's mind before it is mentioned to some able children.

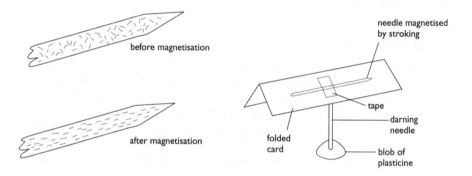

Figure 8.5 A magnetic compass

Magnetic materials

Magnets attract a limited range of materials: steel, iron, nickel and cobalt are attracted by magnets. Some stainless steels are not magnetic because they are alloys containing high proportions of non-magnetic materials. British 1p and 2p coins have been made from steel since the mid-1980s. Prior to that they were made from a non-magnetic alloy.

Electromagnetism

If a single piece of wire is connected to the two ends of a cell, the electric current flowing through it creates a magnetic field. This effect can be detected using a small magnetic compass placed on or very close to the wire. Wire can be wound round and round a tube to form a solenoid. This magnifies the magnetic effect of the single wire. Solenoids are used to make magnets. If a piece of unmagnetised steel is placed in one and a sufficiently large current is passed through the wire coil, the steel becomes strongly magnetised.

Bells use electromagnet effect to attract the clapper onto the bell. In the process of moving towards the bell, the clapper switches off the electric current allowing the arm to spring back to the place where the current is immediately switched on again. Motors rotate when the electromagnet, made from many metres of fine wire, is attracted to a curved magnet which is fixed in the motor. The polarity of the coil is flipped between north and south, causing it to be attracted to different poles of the curved magnet, thus continuing its rotation.

A SUMMARY OF **KEY POINTS**

> **All materials are made of particles which include electrons.**

> **In conductors, electrons can move freely.**

> **Resistance measures the difficulty of electricity moving through a material.**

> **Current is the same throughout a series circuit.**

> **In a parallel circuit, current will flow most easily through the paths of least resistance.**

> **A bulb with a high resistance cuts down the flow of electricity.**

> **Voltage is a measure of the work a current can do.**

> **The voltage of a current leaving a resistor is less than the voltage entering.**

> **Filaments and heating wires glow because of the collisions between the flowing electrons and the fixed atoms.**

> **Power is a measure of the rate at which electricity is used.**

> **Like poles of a magnet repel and unlike poles attract.**

> **An electrical current produces a magnetic effect.**

M-LEVEL EXTENSION > > > > M-LEVEL EXTENSION > > > >

Reflect on the practical task on page 111, where you demonstrated electrical flow round different circuits by getting the children to move around obstacles in the classroom. This is not only an effective way of explaining a difficult abstract concept to children in a more concrete way, but it is also an example of a kinaesthetic activity that will suit the preferred learning style of a significant number of the children in your class. Consider how you could develop similar activities to demonstrate other key scientific concepts that children have found difficult to understand.

REFERENCES REFERENCES **REFERENCES** REFERENCES REFERENCES

Asoko, H. (2002) Developing conceptual understanding in primary science. *Cambridge Journal of Education*, 32(2).

Asoko, H. and De Boo, M. (2001) *Analogies and illustrations: representing ideas in primary science*. Hatfield: ASE.

Chang, H-P, Chen, J-Y, Guo, C-J, Chen, C-C, Chang, C-Y, Lin, S-H, Su, W-J, Lain, K-D, Hsu, S-Y, Lin, J-L, Chen, C-C, Cheng, Y-T, Wang, L-S and Tseng, Y-T (2007) Investigating primary and secondary students' learning of physics concepts in Taiwan. *International Journal of Science Education*, 29(4), 465–82.

Kibble, B. (2002) How do you picture electricity? *Primary Science Review*, 74, 28–30.

Mulhall, P., McKittrick, B. and Gunstone, R. (2001) A perspective on the resolution of confusions in the teaching of electricity. *Research in Science Education*, 31, 575–87.

Osborne, R. and Freyberg, P. (1985) *Learning in Science: the Implications of Children's Science*. Auckland: Heinemann.

FURTHER READING FURTHER READING **FURTHER READING** FURTHER READING

Asoko, H. and de Boo, M. (2009) *Representing Ideas in Science: Analogies and Illustrations*. Hatfield: Association for Science Education.

DfE (2011) *Teachers' Standards*. Available at www.education.gov.uk/publications.

Harlen, W. and Qualter, A. (2009) *The teaching of Science in Primary Schools*, 5th edn. London: David Fulton.

Ofsted (2011) *Successful Science*. London: Ofsted.

9
Energy

Curriculum context

National Curriculum programmes of study

At Key Stage 1, children should be taught that pushes and pulls are examples of forces.

At Key Stage 2, children should be taught about the role of the leaf in producing new material for growth, about the feeding relationships shown by food chains, how nearly all food chains start with a green plant, that electrical circuits require a power supply, and that changing the components in a circuit can make the bulbs brighter or dimmer.

Early Years Foundation Stage

Children must be supported in developing the knowledge, skills and understanding that help them to make sense of the world. Their learning must be supported through offering opportunities for them to undertake practical 'experiments'. By the end of the EYFS, children should:

- investigate objects and materials by using all of their senses as appropriate;
- ask questions about why things happen and how things work;
- find out about and identify the uses of everyday technology;
- observe, find out about and identify features in the place they live and the natural world.

Introduction

Without energy nothing happens. Energy is the capacity to do work. Work is done when objects are moved or heated. Energy can be converted from one form to another such as when coal or oil is burnt in a power station. The heat produced by this process is used to make steam which turns a generator. This mechanical energy is converted into electrical energy. Fuels and food are sources of energy. This energy can be traced back to the plants, including the myriad single-celled plants, which use solar energy to create their structures including leaves, stems, roots and fruits.

Energy is an area which is fraught with conceptual difficulties. It wasn't until the nineteenth century that scientists began to realise that there were several distinct forms of energy. Today, the language we use in everyday talk gets in the way of a deeper understanding of energy. Scientists, for instance, believe that the amount of energy in the universe is constant and it is simply changed between different forms. Energy is not used up, it is simply converted. However, every child knows that you can run out of energy and many adults are aware that the planet's energy supplies are limited. There is little doubt that as an abstract idea energy is difficult to understand. Trying to help learners make sense of these two views about energy is not an easy job and that, to some extent, is why energy does not figure prominently in the National Curriculum for primary schools.

Although there is no individual section of the National Curriculum at Key Stages 1 or 2 headed 'energy', the ideas of energy permeate a great deal of the science curriculum. As a result of this, primary school teachers focus on the effects of energy and rarely study energy itself. For instance, the role of light in the growth of plants is explored as is the need for food for activity in people. The effect of burning materials leading to the formation of new materials is fundamental to understanding the way in which fuels and food are converted into other types of energy. Therefore, trainees are expected to understand some of the fundamental ideas about energy and see how they apply to the ideas they teach in schools.

Energy

Fuel in the form of petrol gives cars the capacity to do work. A car engine does work on the wheels to make the car move. Electrical energy from a battery does work on the filament of a bulb to make it glow. The gravitational pull of the Earth does work on a cyclist as they roll down a hill. A stretched catapult elastic does work on the ball of paper to make it move. A moving hammer does work on a nail to push it into a piece of wood. This notion that energy does work is fundamental to understanding how energy conversions work.

An Englishman, James Joule, did the most to clarify early thinking about energy. He built on observations that, when boring into metal with a drill, a great deal of heat was produced. Joule used apparatus which allowed him to accurately measure the rise in the temperature of water produced by the friction of a machine. Energy is measured in joules (J). This is the same unit as is used to measure work (see below).

*For more
information on
forces, see
Chapter 10.*

Force

A force is a push or a pull. To exert a force you must have energy. Forces can:

- stretch or squash objects;
- change the speed of an object;
- alter the direction of an object's movement.

Examples of force include **gravity, friction**, magnetism and **air resistance**. The unit in which forces are measured is newtons (N). An average woman has a **mass** of approximately 60kg. Standing on a weighing scale, this mass is pulled by gravity with a force of approximately 600N. A car's engine might exert a force of 4000N. The friction from a car's brakes may apply a force of 5000N on a moving car to slow it down.

However, there are many cases where a force is applied yet there is no motion. Even though a great deal of force might be applied to an impossibly heavy piano or a stuck screw, if it does not move then no work has been done (see next section).

Work

Work is done when an object moves, electricity flows or an object is heated. In moving objects, one joule is defined as the work done when a force of 1N moves a distance of 1 metre.

$$1 \text{ joule} = 1 \text{ newton} \times 1 \text{ metre}$$

Here are some examples.

- A person pushing a supermarket trolley who pushes it with a force of 40 newtons for a distance of 20 metres is doing 800J of work on the trolley.
    work done = force x distance moved
    work done = 40N x 20m = 800J

- A helicopter lifting two people weighing 2000N a height of 20 metres off the deck of a sinking ship is doing 40 000J of work on the people (see Figure 9.1).
    work done = 2000N x 20m = 40 000J

PRACTICAL TASK PRACTICAL TASK **PRACTICAL TASK** PRACTICAL TASK

Find out what you weigh in newtons. Calculate the height of a flight of stairs. Calculate the energy you expend raising your body up the stairs. Look at the kilojoules of energy in 100g of chocolate. Figure out what fraction of a bar of chocolate you need to eat to replace the energy.

In electrical energy one joule of work is done when one watt of power acts for one second.

*For more
information on
electrical
potential, see
the section on
power in
Chaper 8.*

The work done on a 3kW fire in one hour is:

watts	×	seconds in hour		=	work done (joules)
3000	×	3600		=	10 800 000 joules

Figure 9.1 Work – 40 000J

The work done on a 10 watt energy-saving bulb in one hour is 36 000 joules. The work done on a 100 watt bulb in one hour is ten times that: 360 000 joules.

Forms of energy

In order to do work, energy has to be converted from one form to another. The energy to push a trolley comes from food. The energy to lift people off the deck of a ship comes from diesel fuel. The energy needed to make a torch bulb light comes from the chemical reactions which take place in a battery. Energy to do work comes in various forms.

For more information on heat, see Chapter 6 and also the section on heat energy later in this chapter.

- *Kinetic energy* is the energy of motion. All moving objects have kinetic energy. The heavier and faster the moving object the more energy it has. A heavy lorry moving slowly will do a great deal of damage, or work, if it hits something. A small bullet will do damage since it is travelling very quickly.
- *Chemical energy* is the energy associated with food and fuels. Food provides us with energy when it is digested inside the body. Petrol provides energy when it is burnt inside a car's engine. Chemical changes always produce new substances. Chemical changes can often be reversed. For instance, carbon and oxygen combine to produce carbon dioxide when a fuel burns or a living thing respires. This process releases energy. When carbon dioxide is broken down into atoms of carbon and oxygen, as happens in photosynthesis, an input of energy is needed.
- *Gravitational potential energy* is the energy of position of an object. A model car at the top of a ramp has potential energy as gravity pulls it down the slope. Gravity does work on falling objects. We can calculate the potential energy of an object if we know its mass and the distance it is raised. The greater the height and the mass the more potential energy an object has.
- *Heat energy* is the energy of hot objects. A splash of hot fat has a high temperature but relatively little heat energy compared with a large hand-hot radiator. Temperature and heat are different quantities. Temperature is a measure of hotness and can be measured with a thermometer. Heat, on the other hand, can be regarded as a form of energy.

For more information on electricity, see Chapter 8.

For more information on sound, see Chapter 12.

For more information on light, see Chapter 11.

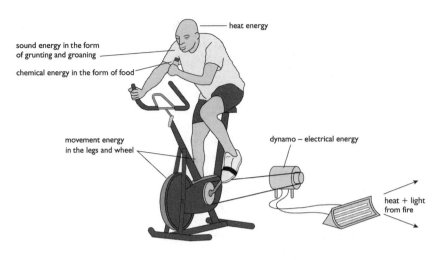

Figure 9.2 Some energy changes in a self-sufficient gymnasium

- *Strain energy* is the stored energy of an elastic band, spring or a bent piece of wood. A bow has strain energy which it transfers to the arrow when it is released. The amount of energy stored in a spring can be calculated by working out the amount of work which has gone into stretching a spring.
- *Electrical potential energy* is the ability of an electric current to do work. A battery makes electrical energy from a chemical reaction.
- *Sound energy* is a form of kinetic energy. Sounds make objects and materials vibrate.
- *Light energy* is a form of electromagnetic radiation. Light energy from the sun can be converted directly into electricity in a solar cell.
- *Nuclear energy* is released when a neutron hits the nucleus of a uranium 235 atom. The atom splits into two parts and smaller particles, releasing a huge amount of energy as it does so. This process is called nuclear fission.

PRACTICAL TASK PRACTICAL TASK **PRACTICAL TASK** PRACTICAL TASK

Set up a small ramp and a toy car. Let the car roll down the ramp and crash into a block of wooden or plastic bricks. What happens to the distance the car can move the bricks when it is weighted with plasticine? Explain it in terms of the energy of the car. Experiment with changing the height of the ramp. Why, in terms of energy, does the car travel faster off a high ramp than off a low ramp?

Energy and fuel

Food and fuel do not contain energy as such. The energy associated with food and fuel is released when the fuel or food is combined with oxygen. Think of a camping gas stove. The fuel is methane gas, the molecules of which are made from atoms of hydrogen and carbon.

For more information on bonding, see the sections on ionic and covalent bonding in Chapter 6.

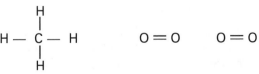

methane molecule   oxygen molecules

For more information on burning, see the section 'What about burning?' in Chapter 7.

When the methane gas is burnt, energy, in the form of heat, is released. This happens when the atoms of methane fuel combine with oxygen. The resulting new materials are water and carbon dioxide:

- water (H_2O) is formed from combining the hydrogen atoms of the gas with oxygen;
- carbon dioxide (CO_2) is formed from combining the carbon atoms of the gas with oxygen.

The new bonds between the atoms of carbon and oxygen in the carbon dioxide and between the hydrogen and oxygen in the water molecules are strong. They are considerably stronger than the bonds between the oxygen atoms in the oxygen molecule, and the carbon and hydrogen atoms in the methane molecule. The energy released when these new stronger bonds are formed is heat. The energy needed to break these new bonds is much greater than that needed to break the weak bonds in the oxygen and methane molecules that they replace.

So far, we have considered only fuels which are burnt. Let us now look at digestion. Glucose molecules have the chemical formula $C_6H_{12}O_6$. In the first part of digestion, the molecule of glucose is pulled apart and this requires an energy input. The energy associated with the sugar is released when the carbon and hydrogen combine with oxygen to produce carbon dioxide and water.

Fuels for generating electricity

There are three main classes of fuel which can be used to generate electricity.

1. Fossil fuels such as gas, oil and coal.
2. Renewable fuels such as wood.
3. Nuclear fuel such as uranium.

RESEARCH SUMMARY RESEARCH SUMMARY RESEARCH SUMMARY **RESEARCH SUMMARY**

Research into the idea that energy is contained in fuels is reported in Littledyke *et al.* (2000). It found that many people believed that energy is released when the bonds between molecules are broken. It argued that the conservation of matter was one of the most important aspects of understanding energy from fuels and food. The key question was:

Suppose we trap all the fumes from the exhaust pipe of a car and were then able to weigh them – we would be able to compare the amount of petrol used with the amount of exhaust collected at the end. Would the mass of exhaust be:

(a) much larger than the petrol?

(b) about the same as the mass of the petrol?

(c) much smaller than the petrol?

Scientists would answer (a) as the process of burning fuel is constructive because the atoms of the fuel are combined with oxygen. It was found that many people believe (incorrectly) that the amount of the exhaust will be smaller than the petrol.

The study also surveyed the stages of people's understanding of the conservation of matter. Young children believe that a crushed biscuit has less mass than a complete biscuit but they develop an understanding of conservation of mass at the end of primary school. However, it was found that many adults did not think of gases as having mass. Many people think of gases, such as air, as weightless. The research argues that the expression 'fuels contain energy' is most misleading and argues for teaching that energy is released from the *fuel–oxygen system*.

The bonds between the atoms of the oil, coal or gas are relatively weak. When they are burnt the atoms of the fuel combine with oxygen to make new substances which are held together with much stronger bonds. In the process of forming these stronger bonds heat is released. This heat is used to boil water and the resulting steam is employed to turn large propellers called turbines. In turn, these turbines turn electrical generators. In many senses **fossil fuels** are stored sunshine. In photosynthesis the energy of the Sun is used to stop the atoms of oxygen, hydrogen and carbon from combining. The rapid burial of fossil fuels away from air means that when they are dug from the ground the reaction between oxygen and the fuel is yet to take place.

For more information on energy in biological processes, see the section on leaves and photosynthesis in Chapter 2.

The process of releasing the energy from the fuel–oxygen system in renewable materials such as wood points out an important aspect of energy release. If wood, instead of being burnt soon after it is cut down, is allowed to rot, the wood will combine with oxygen through the process of decay. The new materials, carbon dioxide and water, are useless as fuels since they require large inputs of energy to break the strong bonds between the carbon and oxygen and between the hydrogen and oxygen. Fossil fuels are available to us millions of years after their formation because oxygen has not been allowed to combine with the plant or animal material.

Heat energy is released from radioactive fuels such as uranium when the nucleus of a uranium atom is bombarded with particles called neutrons, releasing huge amounts of energy when it splits to form other elements. The forces in the atom itself are much larger than the forces between atoms in molecules.

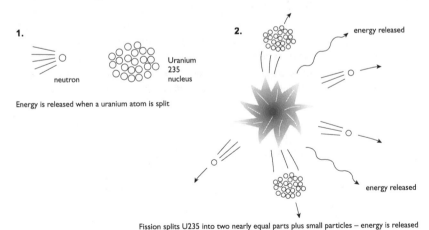

Figure 9.3 Heat energy from nuclear fuel

In addition to these fuels there are other energy sources used to generate electricity.

- *Tidal power* – the motion of tides coming in and out of estuaries can be used to drive turbines. The tides are driven by the gravitational attraction of the Sun and the Moon.
- *Hydroelectric power* – the potential energy of the water is used to turn generators. The higher the drop and the greater the drop, the more energy can be generated.
- *Wind power* – moving air can be used to turn propellers. The energy of the wind can be traced back to the way in which the Sun's energy heats different parts of the Earth to a greater or lesser extent.
- *Wave power* – moving air creates friction across the surface of water and makes waves which can be used to turn generators.
- *Geothermal power* – the interior of the Earth is very hot. Where this heat is close to the surface it can be used to heat water which in turn is used to drive generators. Iceland has huge quantities of heat energy close to the surface and can generate electricity at very low prices.
- *Solar power* – can directly produce electricity using photoelectric cells.

Chemical changes produce electricity

Luigi Galvani noted, in 1791, that a dead frog's leg twitched when it was being dissected with metal instruments. His compatriot, Alessandro Volta, tested different combinations of metals before making the first electric battery from a stack of zinc and copper discs. The discs were separated by cloth soaked in a salt solution. When the top and bottom of the battery were connected by a wire, electricity flowed. Many torch cells today are made using carbon and zinc as the two terminals separated by a chemical paste. Electricity flows from the zinc to the carbon.

RESEARCH SUMMARY RESEARCH SUMMARY **RESEARCH SUMMARY RESEARCH SUMMARY**

Lee (2007) interviewed groups of primary school children and asked about their concepts of electrical batteries. Among the conceptions expressed they found children who considered electricity as material and the battery as a container. Some children thought of the battery as being similar to a petrol tank, with the fuel being poured in. For some children the electricity was stored in the rod in the centre of the battery. Lee suggests that this idea may arise from the children's notion of a carbon rod in the middle of the battery without knowing that the electricity is produced by the reaction inside the battery. The concept of electrical current is often introduced to children using the analogy of water flow. In Lee's results, 71 per cent thought of electricity as water in a reservoir. Another analogy would be energetic little creatures.

For more information on batteries, see the section on cells and batteries in Chapter 8.

Figure 9.4 Understanding energy transfer through the analogy of 'little creatures'

Energy transfers in biological processes

At the base of all food chains, apart from deep-sea geothermal vent communities, are green plants which use the energy from the Sun to drive the chemical reactions of photosynthesis. It is green plants' ability to make starch and sugar from carbon dioxide and water which distinguishes them from all other living things. The complex molecules constructed by plants provide the food needed by animals to fuel their life processes.

In the process of photosynthesis the molecules of carbon dioxide and water are pulled apart using the energy of sunlight. As we have already noted, a great deal of energy is needed to split up these simple stable molecules. Splitting the strong bonds of water and carbon dioxide and forming them into more complex sugars and starches allows energy to be released when the same simple strong bonds are recreated during respiration.

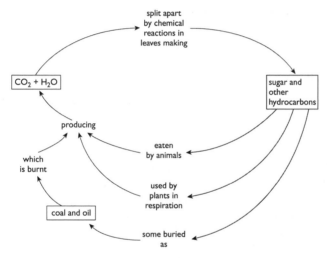

Figure 9.5 Part of the energy cycle

REFLECTIVE TASK

When thinking about changes to materials that are easy or difficult to reverse, remember that nature can reverse apparently one-way reactions such as burning paper. It is impossible for us to turn ash into paper, but for trees this is part of what they do. Carbon dioxide is released by burning and this gas is used in photosynthesis to make more wood which can eventually be turned into paper.

Energy flows within an ecosystem are complex. The energy from the Sun is used by plants to photosynthesise. Some of the plant food is used directly by birds, herbivorous animals and small arthropods such as insects and mites. These small animals are in turn preyed on by carnivores. Note from the figures below that relatively little of the Sun's energy is used by the plants and a tiny proportion of it makes its way to predators. It is probably better to think of plants and the bodies of prey animals as fuel which is passed on up the food chain.

For more information on energy efficiency in a food chain, see Chapter 5.

A typical meadow, for instance, may have the following energy flows:

Energy from Sun	=	6.3×10^6 kJ/m^2 (6 300 000kJ per square metre)
Absorbed by plants	=	10 000 kJ/m^2
Eaten by birds	=	300 kJ/m^2
Herbivorous animals	=	500 kJ/m^2
Herbivorous arthropods	=	1 200 kJ/m^2
Growth of birds	=	3 kJ/m^2
Growth of animals	=	5 kJ/m^2
Growth of arthropods	=	190 kJ/m^2

Converting energy in our bodies

For more information on digestion, see the section on digestive systems in Chapter 3.

Digestion is the process of breaking down food in the gut. Large molecules, such as those of which fats and proteins are made, are broken down into smaller food molecules. Notice that this process requires an input of energy since it involves breaking the bonds between molecules.

Respiration is the process of combining the carbon and hydrogen from simple foods such as glucose (a simple type of sugar) with oxygen. This process takes place in nearly every cell in the body:

$$C_6H_{12}O_6 \; + \; 6O_2 \longrightarrow 6CO_2 \; + \; 6H_2O \; + \; energy$$

For more information on respiration, see the sections on respiration in Chapters 2 and 3.

We use food to fuel our bodily functions. The energy released during respiration is used for growth and converted into heat and movement. In the process of respiration the energy value of food is released slowly. The smoothness of the energy release from food is controlled by a very complex series of reactions which breaks down the glucose into its component atoms before oxidising the carbon. In experiments, mice have been fed sugar ($C_6H_{12}O_6$) which contains traces of radioactivity. A short time after eating this the mice were found to be breathing out carbon dioxide which was radioactive. The carbon in the exhaled air came from the carbon in the sugar thus showing that it is the same atoms of carbon being both taken into the body and then breathed out after undergoing chemical change in the body.

The process of burning is much more sudden. When a piece of food such as a peanut is burnt, carbon dioxide and water are produced and heat energy is given off. A handful of peanuts could produce enough energy, if it was all released suddenly, to raise a person's body temperature by 10°C.

A person's body requires different amounts of fuel to perform functions:

	kJ used per hour
sleeping	270
sitting	420
standing	440
walking	850
running	1500
very hard exercise	2500

Make an educated guess at the amount of energy you expend during the day. Express it in the form of kilojoules. Compare it with the energy input in the form of food.

Food labels give the energy value of foods. For instance:

100g	cornflakes	1560kJ
100g	Cheerios	1588kJ
100g	Shredded Wheat	1410kJ
100g	baked beans	300kJ
100g	chocolate	2195kJ

All the biological processes associated with energy described so far have assumed that there is oxygen present. Many micro-organisms respire without the need for oxygen. Yeast, for instance, breaks down sugar into alcohol and carbon dioxide. It can do the same to carbohydrates in bread dough. Bacteria can respire in the absence of oxygen in sewage farms and in swamps. This anaerobic respiration is a vital part of the carbon cycle since decay by micro-organisms releases carbon dioxide into the atmosphere for plants' use in photosynthesis.

Heat energy

If you lick the back of your hand then blow gently, the damp patch feels cool. This cooling happens because to change the liquid water into water vapour requires the input of heat energy. The heat comes from the hand itself. When you emerge from a swimming pool or the sea you feel cold until you have dried your skin. This cooling effect is even stronger if there is a breeze blowing. Again, the energy for evaporation comes from the body.

The cooling effect of sweat evaporation helps to keep us cool in hot weather. In very humid conditions, when the air is already saturated with water vapour, sweat cannot evaporate off the skin so we get uncomfortably hot and damp.

A pan containing frozen soup or stew takes time to defrost. So long as it is stirred the liquid soup does not increase markedly in heat until all the frozen pieces have melted. This happens because most of the heat input is used to break the bonds between the molecules in the solid. Only when all the solid has liquefied does the temperature of the liquid rise rapidly. Once boiling point is reached the rate of evaporation exactly balances the input of heat. The more heat energy supplied to a pan, the more rapidly the liquid evaporates.

Before fridges were common people used to keep bottles of milk fresh by placing them in a bowl of water and draping a damp cloth over the bottle. The bottles were kept cool, not by the coolness of the water, but by the evaporation of the water from the cloth.

Heat energy is transferred by three processes.

1. Conduction.

2. Convection.
3. Radiation.

In conduction heat energy is passed through a solid from the hot to the cooler parts. The molecules of a hot material have more energy than the molecules of a cool material. This energy takes the form of movement – kinetic energy. In a solid the molecules close to the source of heat vibrate and pass this energy on to neighbouring molecules. Good heat conductors include metals. Poor heat conductors include most non-metals, for example asbestos.

Convection occurs in liquids or gases. In a pan, the water next to the heat source expands because the molecules are more energetic and require more space. The warm water rises because it is less dense than the surrounding water.

For more information on light, see Chapter 11.

The Sun's heat reaches the Earth through the vacuum of space. Heat energy that travels in this way is radiant heat or infrared radiation. Radiant heat can be reflected by white clothing or shiny materials. This effect is used when people wear white clothes in hot countries or where shiny survival blankets are used to reflect back the body heat of a person. On the other hand, black or matt surfaces absorb heat radiation and cause the object to heat up.

Energy changes

Energy cannot be created or destroyed. Energy can only change in form.

• A ball which is thrown up into the air and caught again involves: solar energy – chemical energy associated with food – kinetic energy of the moving arm – kinetic energy of the ball during its flight upwards – energy of position (gravitational potential energy) increasing towards the top of the flight – kinetic energy increasing as the ball falls.

• A bungee jumper jumping off a bridge involves: gravitational potential energy on the bridge – increasing kinetic energy as they fall – increasing elastic strain potential energy in the elastic – increasing kinetic energy and energy of position as they spring upwards – energy dissipated as they hang motionless before being helped down.

• A loudspeaker producing loud music involves: solar energy – chemical energy associated with coal (the plants in the Carboniferous period used solar energy to make leaves and stems which are preserved as coal) – heat (as the coal is burnt) – kinetic (as the steam turns the generators) – electrical – sound – heat. In the last part the sound energy is spread out until it is no longer detectable. It is also absorbed by soft surfaces.

RESEARCH SUMMARY RESEARCH SUMMARY RESEARCH SUMMARY **RESEARCH SUMMARY**

Primary Science Review is an excellent source of ideas and the results of research. Issue 87 (March/April 2005) contained a number of articles about energy and energy changes:

Barton (2005) gave an account of alternative energy sources.

Raffan (2005) explained how children can become involved in investigations using simple turbines.

Psycharis and Daflos (2005) described the effects of ICT modelling on children's learning about solar energy.

Lowe (2005) analysed the way pop bottle rockets can be used to teach about energy.

Efficiency of energy transfers

Energy transfers are never 100 per cent efficient in the way intended by people. When we pass electricity through a light bulb we aim to produce light, but if we touch a bulb we know that it is very hot. This heat is wasted energy. When we use the energy associated with petrol to move a car, much of the energy is not converted into movement; instead some of it is converted into noise and heat.

For more information on human energy transfers, see the section on converting energy in our bodies earlier in this chapter.

A cyclist hopes to use all of their energy to work against air resistance – instead, most of the energy of the food is wasted:

chemical energy in food → maintaining body temperature
→ work against air resistance
→ heat loss from the body
→ noise of the cycle
→ heat of moving parts of the cycle

A refrigerator takes heat energy from inside the fridge and discharges it into the room which gets warmer. In the cooling coils of a fridge a liquid is allowed to expand and evaporate thus absorbing heat from the body of the fridge. The gas is then compressed outside the fridge where the absorbed heat is released.

A SUMMARY OF **KEY POINTS**

> **Energy is the capacity to do work.**

> **Work is done when an object moves, electricity flows or an object is heated.**

> **Energy can be in a number of different forms.**

> **Energy can be converted from one form to another.**

> **Energy cannot be created or destroyed.**

> **Food and fuel do not contain energy but energy is released when the food or fuel is combined with oxygen.**

> **Energy is released from nuclear material when nuclei are split.**

> **Chemical changes produce electricity.**

> **Food pyramids summarise energy transfers in biological systems.**

> **Heat energy moves from hotter objects to colder.**

M-LEVEL EXTENSION > > > > M-LEVEL EXTENSION > > > >

Think about how children's learning about energy in science could benefit from being set in a cross-curricular context. Would it be possible to link this aspect of science with looking at how humans use energy in PE, and food as a fuel and healthy eating in personal, social and health education. What about a study of alternative energy sources in geography and role-play work in drama on the arguments for and against the siting of a new wind farm in the local area?

REFERENCES REFERENCES **REFERENCES** REFERENCES REFERENCES

Barton, L. (2005) Renewable energy for the next generation. *Primary Science Review*. 87, 4–7.

Lee, Shyan-Jer (2007) Exploring students' understanding concerning batteries – theories and practices. *Internationl Journal of Science Education*, 29(4), 497–516.

Littledyke, M., Lakin, E. and Ross, K. (2000) *Science Knowledge and the Environment*. London: Fulton.

Lowe, G. (2005) Making pop bottle rockets. *Primary Science Review*, 87, 23–25.

Psycharis, S. and Daflos, A. (2005) Solar energy: a journey from the Sun to the Earth. *Primary Science Review*, 87, 17–19.

Raffan, D. (2005) What gets the turbine spinning? *Primary Science Review*, 87, 8–11.

FURTHER READING FURTHER READING **FURTHER READING** FURTHER READING

Breithaupt, J. (2001) *Physics*. Cheltenham: Nelson Thornes.

DfE (2011) *Teachers' Standards*. Available at www.education.gov.uk/publications.

Lewis, R. and Evans, W. (1997) *Chemistry*. London: Macmillan.

Sutton, J. (1998) *Biology*. London: Macmillan.

10
Forces and motion

Curriculum context

National Curriculum programmes of study

At Key Stage 1, children should be taught to find out about, and describe, the movement of familiar things (for example, cars going faster, slowing down, changing direction), that both pushes and pulls are examples of forces, and to recognise that when things speed up, slow down or change direction, there is a cause (for example, a push or a pull).

At Key Stage 2, children should be taught about types of force: about the forces of attraction and repulsion between magnets, and about the forces of attraction between magnets and magnetic materials; that objects are pulled downwards because of the gravitational attraction between them and the Earth; about friction, including air resistance, as a force that slows moving objects and may prevent objects from starting to move; that when objects (for example, a spring, a table) are pushed or pulled, an opposing pull or push can be felt; and how to measure forces and identify the direction in which they act.

Early Years Foundation Stage

Children must be supported in developing the knowledge, skills and understanding that help them to make sense of the world. By the end of the EYFS, children should:

- investigate objects and materials by using all of their senses as appropriate;
- find out about, and identify, some features of events they observe;
- look closely at similarities, differences, patterns and change;
- ask questions about why things happen and how things work.

Introduction

We can get into deep water when trying to understand the concept of force and its relationship to movement. Our intuitive understanding and our everyday use of non-scientific language can cause some confusion. Having said that, a study of forces is fundamental to appreciating many of the key ideas in science. The interplay between force and movement dominates our everyday experiences, whether we recognise this or not. This science topic is fascinating and touches on a wide range of phenomena. It is not something which can be covered in a single school topic. More success is gained from building up children's understanding in small doses.

If you have wondered why some things float and others don't, or how friction and gravity behave, then you need to know something about forces. If you have ever struggled with a supermarket trolley or driven on a slippery road, then you will be interested in the way that forces can control movement. This chapter sets out to clear up some of the more common confusions and relate the science of forces and movement to everyday life.

RESEARCH SUMMARY RESEARCH SUMMARY RESEARCH SUMMARY **RESEARCH SUMMARY**

Harlen and Qualter (2004) referred to the research into children's understanding over the years. Many of these investigated the ideas about forces, revealing some recurring misconceptions and difficulties in teaching this topic. Children may have a number of ideas that are not the same as those that scientists hold about forces. Among the alternative ideas held by children, and some adults, are:

1. **The human-centred view**
   Human or animal attributes may be given to inanimate objects to explain what happens, e.g. 'a spring uncoils because it wants to get back to its original shape'. Young children, unsurprisingly, may recognise only those forces that they apply themselves by pushing or pulling objects.

2. **Forces must have some obvious effects, mainly movement**
   This is the most common misunderstanding associated with forces. Most often forces are perceived only when there is some associated movement, e.g. 'Friction is stopping the box so it is not a force', 'The ground does not push up on a car'. So it may be difficult to teach about forces which apply to structures of stationary objects.

3. **Force is something contained in objects or people**
   People may think of a force as being given to an object when it is made to move, e.g. 'The force on the ball is what I have put into it by whacking it'.

Harlen (2007) reviewed the work of the SPACE project which researched children's and teachers' ideas about scientific concepts. She reported that some teachers had a similar incorrect understanding of scientific phenomena as their pupils. Rather than a council of despair, Harlen reported on the positive results of involving teachers in exploration using the same materials and equipment they would use with children. After doing this they were in a position where they were able to work alongside their pupils in an exciting learning environment.

Excellent starting points for this learning can be found in Naylor and Keogh (2004) who advocate a wide range of approaches to learning using drawings as the focus for discussion about scientific ideas. What do you think is the correct answer to 'I wonder what it feels like if an elephant steps on your toes – would it be painful, crushing, inconvenient or slightly painful?' Naylor and Keogh also asked readers to speculate about the path of stones dropped from the hands of giants 500 miles high who are standing at different points on the Earth. How will gravity work on them? Is it always downwards?

Wilson (2001) also discussed the force of gravity and the relationship between mass and weight. She tackled the idea that if we could remove air resistance then all objects would fall at the same rate. This was shown when an astronaut standing on the Moon dropped a large white feather and a hammer simultaneously. Because there was no air resistance they fell at the same speed.

The difference between force and movement

Let us begin by separating in our minds the idea of force and that of movement. It is easy to get these mixed up. The language and symbols we use to describe force and movement do not help. The National Curriculum for primary science and many other sources use 'pushes and pulls' as an initial, simplified way of describing forces. Unfortunately, these words most often suggest movement of some kind – we rarely use 'push' or 'pull' when something is motionless, and yet forces act on things which are still as well as on those which are moving. Equally confusing is our use of arrows to show both force and movement, especially in such things as cartoon illustrations. This is true in comic strips as well as in scientific or engineering diagrams in which arrows sometimes show the way something is moving and often indicate how fast this is.

Movement describes how an object changes its position. We can think of five kinds of 'movement'.

1. No movement – remaining still.
2. Getting faster and faster – acceleration.
3. Getting slower and slower – deceleration.
4. Movement which is not in a straight line – a change of direction.
5. Moving at a constant speed in a straight line – constant velocity.

Forces are invisible but can often be described by their effects. Forces can:

- hold things still;
- make things accelerate;
- make things decelerate;
- change the direction of things;
- enable things to continue to move at a constant velocity.

It is possible for things to stay still or move with a constant velocity with no forces acting on them – but this can happen only in outer space, far away from any gravitational pull.

Forces are also needed to change the shape of things. Imagine squeezing a ball of modelling clay or foam rubber. This can conveniently be thought of as *moving* one part of the material from one place to another. When you bend a piece of card you are moving one part of it with a force. When the force is removed the distortion stops.

Examples of everyday forces are:

- pushes or pulls provided by muscles or machines;
- wind or moving water;
- stretched elastic bands or springs (i.e. springy materials);
- buoyancy in water (or air);
- friction between solid surfaces;
- air or water resistance;
- reaction forces (explained later in this chapter);
- magnetism;
- gravity;
- forces between molecules.

The last three on the list are examples of forces which do not rely on some form of physical contact with the object that the force is acting on. The others do.

For more information on magnets, see the section on magnetic poles in Chapter 8.

PRACTICAL TASK PRACTICAL TASK PRACTICAL TASK PRACTICAL TASK

Magnets can be used in the classroom as an example of a non-contact force which can 'be held in the hand'. Try these activities yourself before introducing them to children.

1. Take a magnet, a steel ball-bearing and a horizontal table top. One pole of the magnet provides an excellent representation of a force. Use your 'force' to make the stationary ball-bearing move *without touching it*. Watch closely. Does the ball-bearing accelerate? Does the magnet provide a push or a pull?
2. Roll the ball-bearing slowly across the table. Use your 'force' to slow the ball-bearing down without touching it. In which direction does the force work?
3. Roll the ball-bearing again. This time, use your 'force' to make its direction change. In which direction does the force act to make the ball-bearing move in a curve?
4. Try using two different strength magnets to make the ball-bearing move. How many different positions can you hold these in to achieve movement?

Answers

1. The ball-bearing must accelerate (get faster and faster) because it starts from rest. The magnet will be pulling the ball-bearing.
2. The magnetic force must work in the opposite direction to the ball-bearing's movement.
3. The magnetic force must pull towards the inside of the curved path.
4. If the magnets are of different strength and held at the same distance from the ball-bearing, they can be held at any angle to each other to produce movement.

Balanced and unbalanced sets of forces

Forces can occur in groups. If, for instance, two forces are equal, but pull in opposite directions, they will 'cancel each other out' and the object will remain still. These are **balanced forces**. Imagine two children pushing with equal force on either side of an open door. The door will not move. More than two forces can have a balanced effect. Imagine a group of people playing with a Ouija-board with their fingers pressing lightly on the upturned glass. The glass will remain still when the different forces on it, acting in different directions, all balance.

If a pair of forces acting on an object are not in balance, for instance if one is larger than the other or they do not pull in opposite directions, then the object will move. These are described as **unbalanced forces**. Just as before, more than two forces can also be unbalanced.

Speed, velocity, acceleration and deceleration

Before we go on to look more closely at forces, we need to clarify the terms used to describe movement. The **speed** of something is a measure of how far it travels in a set time. So 'miles per hour' describes how many miles would be travelled in one hour. Over a simple journey, the average speed is the distance travelled divided by the time taken. This may result in speeds such as 30 miles per hour (mph), 5 centimetres per second (cm/sec), and so on.

Velocity describes both speed and direction, e.g. a cyclist travelling at 3 metres per second in a northerly direction.

Acceleration describes a changing velocity. This may be a result of something getting faster and faster. A car may accelerate from 0 to 60 mph in 10 seconds or a falling stone may increase its velocity by 10 cm/sec each second (or 10 cm/sec^2). Equally, a *change in direction* would also constitute an acceleration, even if the speed remained constant. This means that something which moves on a curved path at the same speed, such as a ball swinging round on the end of a rope, is accelerating and so requires a force to do so. This force is supplied by the rope. If the rope breaks, the force disappears and the ball will be able to continue only in a straight line (which will be at a tangent to the curve it was travelling along).

The relationship between force and movement

From our definitions above, a force or set of forces is required to make things move in different ways, but also forces can act on things which remain still, as long as the forces are balanced.

It is easy to see that balanced forces acting on an object will result in its staying still. It is much harder to appreciate that balanced forces will enable an object to continue to move at a constant velocity. When skydivers jump from a plane, they begin to accelerate downwards. At this point, the greatest force on each of them is that due to gravity – downwards. As their speed increases, the air resistance (a force acting upwards) becomes greater. Eventually the air resistance will be equal to the force due to gravity and they will have reached their terminal velocity. The skydivers now fall at a constant speed and the forces on them are balanced.

It requires an unbalanced set of forces to act on an object in order to make it accelerate, decelerate or change its direction. If you use the muscles in your arms to push a supermarket trolley away from you, it will initially accelerate from being still to travelling down the aisle. While your arm is in contact with the trolley you are providing an accelerating force. As soon as you let it go, however, the trolley will begin to slow down. This is because a combination of **friction** in the wheels and air resistance will act in the opposite direction to its movement. So forces acting in the opposite direction to movement will produce a **deceleration** resulting in the trolley coming to a halt.

Things get slightly more complicated when you try to push the trolley round a sharp corner! We can imagine the trolley travelling along the curve of a circle. The force required to make this happen acts towards the centre of this circle. This force will be provided (rather awkwardly) by yourself and also friction between the wheels and floor. If someone has dropped a slippery liquid on the supermarket floor, your trolley may not get enough friction force and go careering off in the wrong direction! A similar thing happens when you take a corner in a car and there is ice on the road.

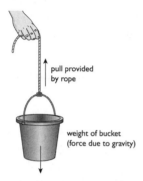

pull provided by rope

weight of bucket (force due to gravity)

Figure 10.1 Forces acting on the bucket only

Thinking about the forces on an object – a bucket on a rope

Yet another source of confusion about forces occurs when we are not clear in our minds about what the forces are acting on. The best way to avoid this is to consider the forces on *one* object at a time. Imagine a hand is holding a rope tied to a bucket. The forces on the bucket will be:

- gravity acting downwards (the weight of the bucket);
- the pull of the rope upwards.

Because these two forces are balanced – equal but acting in opposite directions – the bucket will remain still. They can be represented as in Figure 10.1. Notice the forces are represented by arrows of equal size which show the direction in which the forces are acting.

REFLECTIVE TASK

Look at the diagram in Figure 10.1 and mark the forces acting on the rope only. What can you say about the sizes of these forces on the rope?

Answer

There would be a force acting downwards on the lower end of the rope provided by the bucket, and one acting upwards on the top end of the rope provided by the hand. Both forces should be equal in size to the ones in Figure 10.1.

Thinking about the forces on an object – moving a box

Think of the forces acting in Figure 10.2. The person is pushing the box along the floor. To avoid confusion we will think only of the forces on the box. There are a number of important points to make about this situation.

- The weight of the box is a force provided by gravity.

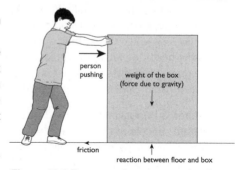

person pushing

weight of the box (force due to gravity)

friction

reaction between floor and box

Figure 10.2 Forces acting on a moving box

- This is balanced by an equal but opposite force provided by the floor called the reaction force. If the floor was not there the box would fall. (If the box was being held in position by a person they would have to push hard upwards just to hold the box still.) Many children find it difficult to recognise the existence of a reaction force.
- The person pushing provides a force on the box from left to right.
- This is resisted by a friction force caused by contact between the floor and the box. If the pushing force is greater than friction the box will accelerate forwards.
- Overall, the four forces are unbalanced enabling the box to move.

For more information on friction, see the section on friction and resistance later in this chapter.

REFLECTIVE TASK

Look at the diagram in Figure 10.2. Think of four forces acting on the *person* before the box moves.

Answer

The person is acted on by these forces.

- Gravity acting downwards – the person's weight.
- A reaction force upwards provided by the floor (equal to the person's weight).
- A 'push back' or reaction force from the box acting through the hand from right to left.
- Friction between the person's shoes and the floor acting from left to right.

Notice that the 'push back' from the box will be equal and opposite to the push from the person on the box. You will often read in books that forces occur in pairs which are equal and opposite. This is what it means, but this can lead to confusion because the two forces are acting on different things.

Measuring force

Forces are measured in newtons (N) after Isaac Newton (at least in the metric world). A force of one newton is quite weak. Hold an apple in your hand and feel it being pulled downwards by gravity with a force of about 1 newton. Push a swing door open with a force of about 4 or 5 newtons. Force can be measured with a force meter or newton meter. These contain springs which stretch in proportion to the force under investigation. Unfortunately, we often measure the mass of an object on the surface of the Earth with a spring balance. This may look the same as a force meter but will have a different set of markings (in grams or kilograms) on the scale. For this reason it is wise to use force meters which look quite different from the spring balances used in the same school.

Reaction forces

We have already mentioned reaction forces as being those which act as a result of another force. Thus, the weight of a book on a table will result in an equal and opposite reaction force being set up by the table, which will enable the book to remain still. The reaction force did not exist before we put the book on the table. The person pushing the box in Figure 10.2 has set up a reaction force provided *by* the box *on* the person.

Children can be helped to understand reaction forces by replacing the table, for instance, with their own hand (which must now push upwards to hold the book still), a strong spring (which appears depressed with a book on it) or perhaps a set of bathroom scales. Reaction forces generally act at right angles to the surface providing them (see Figure 10.4 for an example of this).

Gravity

We often talk about *the* force of gravity but in fact gravitational attraction produces different forces on different objects. Gravitational attraction describes the forces which occur between all objects. The law of gravity states that there is a force of attraction between any two objects which is proportional to their masses and this force gets less as they move further apart. Most objects that we are familiar with are too small in mass to have a noticeable attraction – that is, until we consider the planet Earth (large mass) and anything on its surface. The gravitational force between you and the Earth is called your weight. Gravity acts by pulling you towards the centre of the Earth. You also pull the Earth towards you with the same force! Clearly, the Earth's gravitational force on the book you are reading is different from and much less than that on yourself. The force due to the Earth's gravity on a one kilogram mass is about 10 newtons, on a two kilogram mass about 20 newtons, and so on, because the force of gravity at the surface of the Earth is about 10N/kg.

As we move further from the Earth, the effect of gravity is diminished, though we would have to travel a long way before it became negligible. The apparent weight-lessness of astronauts can be explained by the fact that gravity pulls their spacecraft into a circular orbit. The forward motion of the spacecraft is at right angles to the pull of gravity so it is not pulled down to the Earth's surface.

Galileo's experiment

Gravity has a peculiar effect on falling objects and Galileo (or, some say, one of his contemporaries) set out to demonstrate this by climbing the Leaning Tower of Pisa. Galileo understood that, contrary to our intuitive feelings, heavy objects will fall at exactly the same rate as lighter objects. Thus, two different-sized cannon balls were shown to fall from the top of the tower and land at the bottom at practically the same time. We need to ignore air resistance to understand this, so dropping a feather and a marble, for instance, will not work because the air will slow the feather much more effectively.

The explanation goes something like this. A force is needed to make an object accelerate. A falling object accelerates due to the force of gravity acting on it. A light object will have a relatively smaller accelerating force acting on it. However, a lighter object is relatively easier to get moving. (Consider the difference in getting a car moving by pushing and getting a bicycle moving.) Therefore, a light object will have a small force due to gravity available to make it accelerate downwards but only needs a small force to get moving. A heavy object has a large force on it due to gravity but requires that large force to get going. Then, ignoring air resistance, all objects accelerate downwards due to gravity at the same rate which is about 10 metres per second each second.

There is a wonderful video of an astronaut standing on the Moon dropping a hammer and a feather. Both fall to the ground simultaneously for the reasons outlined above and because there is no air resistance to complicate matters.

The next time the children in your class claim that the heavier toy car will roll down the slope fastest, be aware that this may not reflect the scientist's view. Also consider that the situation will be complicated by many factors, such as the different friction forces in the wheels of the toy and the nature of the surface these roll on, and so on. It *is* a worthwhile investigation, but beware of trying to explain the complex science involved.

Throwing a ball up

We can now bring together some of the ideas discussed so far in a description of a simple action. Imagine throwing a ball across the playground. The hand holding the ball begins to move from rest so it must accelerate in the direction of the throw while it is in contact with the ball. As the ball leaves the hand, however, this accelerating force is removed. All that is left is the **weight** of the ball acting downwards and a little air resistance acting against the direction in which the ball is moving. The forces on the ball as it travels on its journey, then, act in the opposite direction to its movement (although our intuition suggests that there is still a force in the direction of movement!).

In fact, the two forces acting on the ball make it slow down from the time it leaves the hand. Eventually, the ball gets to the top of its trajectory, then begins to accelerate downwards under the influence of the force due to gravity.

Mass and weight

You will see that we have already started to use the ideas of mass and weight. These two things are often confused. The mass of something is a measure of how much matter it contains and is measured in grams or kilograms or pounds. The weight of something is a measure of the pull of gravity on it. It is a force which is measured in newtons. Confusion begins when we talk about our 'weight' in kilograms. What we mean is our mass in kilograms. It just happens that a convenient way of measuring our mass is to stand on a set of scales and measure the pull due to gravity on us, i.e. our weight in newtons. Of course, the scale shows kilograms or stones and pounds directly. This is all possible because the pull of gravity is proportional to our mass on the surface of the Earth.

On the Moon, astronauts retain the same mass as on Earth but their weight is less due to the smaller gravitational attraction between them and the Moon. Hence, we can watch them bouncing around in a quite unusual way!

Friction and air or water resistance

In solids, friction is a force which occurs between two surfaces which may be still or moving. It tends to either prevent movement ever beginning or act against it once it has started. If we stand still on sloping ground, the friction between the soles of our shoes and the ground will prevent slipping. If the surface is wet or muddy we will begin to slide downwards but friction will still be acting against this movement and preventing it becoming too large. On a snow slope, the friction is very small between the toboggan and the snow, so gravity will supply a dominant force downwards.

Friction in solids occurs because of the interaction between the two surfaces in contact with each other. This interaction can be a result of the roughness of the surfaces, the pressure between them or the physical nature of the materials involved. Imagine looking at two surfaces under the microscope. They will appear to have jagged and irregular surfaces which will 'catch' in each other as movement is attempted. This is similar to the bristles on two brushes laid one on top of the other (see Figure 10.3).

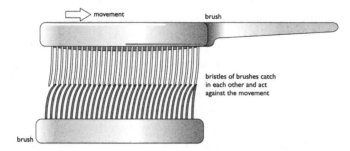

Figure 10.3 Two brushes moving over each other

We can use this model to think about how friction can be increased or overcome. Friction can be increased by:

- having rougher surfaces – more bristles to catch;
- having a greater pressure between the surfaces – this pushes the bristles more into each other;
- changing the nature of the materials involved – for instance, by using rubber we can get more interaction (grip) between the surfaces.

Friction can be reduced or overcome by:

- having smoother surfaces – fewer bristles to catch;
- having less pressure between surfaces – allows the brushes and bristles to separate and interact less;
- using a lubricant such as grease, oil or water – the 'liquid' forms a film between the two surfaces which keeps them apart;
- reducing the area of contact such as by using ball-bearings in a bicycle wheel – this reduces the number of bristles that can catch.

Using slopes in the classroom

Teachers often use an activity which involves objects moving down a sloping surface. These include toy cars, balls or cylinders or shoes (investigating grip). Gravity, reaction forces and friction all come into play in this situation. If the block in Figure 10.4 is not slipping down the slope, the three forces will be balanced. If, however, the part of the force due to gravity (weight), which acts downwards *along the surface of the slope*, can overcome the friction, the block will slip.

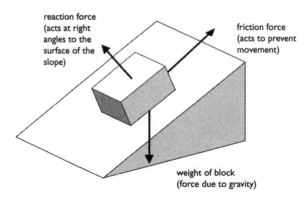

Figure 10.4 Forces acting on a block on a slope

A useful idea for teachers to understand here is that a force such as that due to gravity can be thought of as having two components or parts. Forces due to gravity act vertically downwards but in this example each force can be represented by:

- a part which acts down the slope;
- a part which acts at right angles to this, straight into the slope surface.

It is the first of these which tries to pull the block down the slope. Notice that the reaction force will be the same size but in the opposite direction to the component of the gravitational force which acts into the slope (see Figure 10.5).

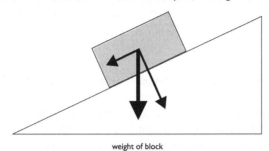

weight of block

Figure 10.5 Any force can be represented by two alternative forces which have exactly the same effect

Measuring friction in the classroom

Often, teachers want children to explore friction by asking them to measure it using a force meter. The usual activity has children pulling a block or a trainer along the table and taking the reading on the newton meter. We might ask if the children really understand the relationship between this reading and the friction being measured. There are a number of points that the teacher should bear in mind.

- The reading on the force meter will vary as the block is pulled along the table making it quite difficult to read.
- The pulling force will equal the friction force only when the block is being pulled at a constant speed. Only at this stage will the forces on the block be balanced.
- The friction force will be acting in the opposite direction to the way the force meter is pulled.
- Just before the block moves, the friction force will be greater than it finally settles down to be, so the children should wait until the block is moving smoothly forwards at a constant speed before taking a reading.

Of course, there are other ways of gauging the friction force, such as increasing the slope of the surface until the block moves.

Floating and sinking

Why things float and sink can be easily explained using our knowledge of forces and movement. Many children never get beyond an appreciation of which materials float and sink in the primary school and yet they could understand much more.

Some adults have memories of learning about density and use this to explain why things float, and indeed this is one way of thinking about the phenomenon, although this is not so appropriate for primary-aged children. A simpler way is to understand that when something floats it *stays still* and when it sinks it is *moving*. We know that the forces on a still object are balanced and those on an object as it begins to move are unbalanced.

With EYFS children, water play is very important. Make it even more useful by restricting from time to time some of the equipment you provide. Cut the bottom off a small drinks bottle. Use the bottle with the screw cap in place as a boat. It floats with great stability with marbles in it. Float the bottle in a calibrated jug to see the change in the volume of water displaced as more and more marbles are added to the boat.

Upthrust

What are the forces on an object in water? We know that gravity provides a down-wards force on all things but, added to this, in water, there is also a force called **upthrust** or **buoyancy**. Upthrust acts upwards on the object but not at any one particular point. Thus, a floating object has its weight acting downwards and upthrust acting upwards and these two forces are balanced. When an object floats, upthrust equals the weight of the object. When an object sinks, its weight will be greater than the upthrust so that it is able to move downwards.

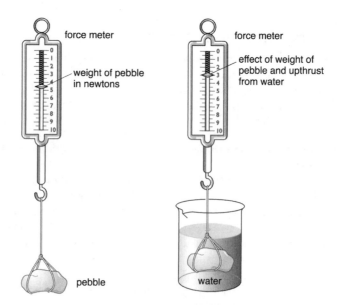

Figure 10.6 Measuring upthrust

Children can easily measure the weight of an object (using a force meter) and just as easily measure the upthrust as it sinks (see Figure 10.6). They dip the object, suspended from the force meter by string, into the water. The new force meter reading now gives the overall downwards pull on the object, which is weight minus upthrust. Of course, if the object floats, then the upthrust is *equal* to the weight of the object.

Incidentally, children can discover that the upthrust will be the same no matter what depth the object is lowered to. Do not confuse upthrust with water **pressure** which increases with depth.

There is a way of thinking how upthrust works which may be of help but which need not be explained to primary children. Imagine a block of steel lowered into some water. The water which it displaces (or pushes to one side) can be thought of as a 'block' of the same volume and shape. This imaginary 'block of water' was previously supported by the surrounding water. This support is now given to the steel block and is called the upthrust on the block. The block of steel will be heavier than the 'block of water' so the upthrust is not sufficient to stop it sinking.

From the description above it should be clear that the upthrust from the water is proportional to the volume of water displaced or pushed aside. This helps us to explain why a steel boat will float while a steel knitting needle will not. The boat shape is designed to push as much water as possible aside, thus increasing the upthrust available. As the heavy steel boat settles in the water, a point is reached when enough water has been pushed aside for the upthrust to equal the weight of the boat – a prerequisite for floating! Children who make plasticine 'boats' will find that their hollow 'dishes' will float when these are able to displace sufficient water.

A SUMMARY OF **KEY POINTS**

> **The effects of forces are all around us and are provided by a variety of factors such as magnetism, gravity, weight, friction and reaction forces.**

> **Forces can act on things which remain still as well as those which move.**

> **There are different kinds of movement which are produced by different combinations of forces.**

> **A balanced set of forces will hold something still or allow it to continue moving at a constant velocity.**

> **An unbalanced set of forces will enable something to accelerate, decelerate or change direction.**

> **Forces are measured in newtons with a force meter.**

> **There is a difference between the mass and the weight of an object.**

> **Friction, including air and water resistance, is a force which either opposes motion or prevents potential motion.**

> **Objects which float are held still by balanced forces – those which sink are able to move downwards due to unbalanced forces.**

M-LEVEL EXTENSION > > > > M-LEVEL EXTENSION > > > >

Consider how you could start to underpin young children's understanding of forces and motion through the guided play activities that you plan in the various areas of your Early Years classroom. How could sand and water play contribute? What about the opportunities provided by the role-play area? Are there any links with other areas of learning, such as Physical Development or Creative Development? How would you then build on this early understanding to ensure progression in children's ideas into Key Stages 1 and 2?

REFERENCES REFERENCES **REFERENCES** REFERENCES REFERENCES

Harlen, W. (2007) The SPACE legacy. *Primary Science Review*, 97, 13–15.

Harlen, W. and Qualter, A. (2004) *The Teaching of Science in Primary Schools*. London: David Fulton.

Naylor, S. and Keogh, B. (2004) *Active Assessment: Thinking, Learning and Assessment in Science*. London: David Fulton.

Wilson, H. (2001) Time to think about forces. *Primary Science Review*, 70, 8–10.

FURTHER READING FURTHER READING **FURTHER READING** FURTHER READING

DfE (2011) *Teachers' Standards*. Available at www.education.gov.uk/publications.

Peacock, G. (2002) *Teaching Science in Primary Schools – A Handbook of Lesson Plans, Knowledge and Teaching Methods*. London: Letts.

Wenham, M. (2005) *Understanding Primary Science – Ideas, Concepts and Explanations.* London: Paul Chapman Publishing. This contains alternative ways of thinking about floating and sinking together with some interesting force diagrams. It tackles action-at-a-distance forces such as magnetism and gravity in a separate chapter.

11

Light

Curriculum context

National Curriculum programmes of study

At Key Stage 1, children should be taught to identify different light sources, including the Sun, and that darkness is the absence of light.

At Key Stage 2, children should be taught that light travels from a source. They should be taught that light cannot pass through some materials, and that this leads to the formation of shadows. They should be taught that light can be reflected from surfaces such as mirrors and polished metals. They should also be taught that we see things only when light from them enters our eyes.

Early Years Foundation Stage

Children must be supported in developing the knowledge, skills and understanding that help them to make sense of the world. By the end of the EYFS, children should:

- investigate objects and materials by using all of their senses as appropriate;
- find out about, and identify, some features of events they observe;

- look closely at similarities, differences, patterns and change;
- ask questions about why things happen and how things work.

Introduction

Light is an essential ingredient of most of human endeavour. Without it we would lose an important means by which we collect information about the world around us. The effects of light can warn us of dangers, enable us to communicate in a variety of ways and provide us with pleasurable experiences.

Most children already know a lot about light and the way in which it behaves. In some instances they will have their own ideas which are different from those of scientists (see below). Teachers will want them to learn more so that they can be safe, can gain pleasure from a greater understanding and can learn to control this phenomenon to the advantage of all.

RESEARCH SUMMARY RESEARCH SUMMARY RESEARCH SUMMARY **RESEARCH SUMMARY**

In her article for *Primary Science Review* No. 64 (2000), Harlen reviewed a wide variety of research into the way children learn about light. She noted the evidence which points to the importance of starting from ideas about light that children have worked out for themselves. However, the conflict between children's ideas and the accepted scientific ideas can sometimes present difficulties. Even young children have a clear idea about primary light sources such as the sun or lamps, but children find the concept of reflection difficult to grasp. Harlen's research showed the importance of first-hand experience of light. She strongly advocated using torches, coloured cellophane and other objects in a dark room or large cupboard. She cautioned that the fact that light travels is not always clear to children and as a consequence ray diagrams are of limited use.

Light comes from a source

Imagine looking at a lighted candle or perhaps a computer screen. These simple everyday activities demonstrate some key ideas about light. Both the candle and the computer screen give out their own light by converting various forms of energy into light energy. If we viewed the candle or monitor in a completely darkened room we would still see them. If, however, we covered our eyes completely, our brain would not be able to register the presence of a source of light at all – we would not *see* anything. The key points here are that:

- light is a form of energy;
- light comes from a source;
- we see things when light enters our eyes and the brain interprets this as 'seeing'.

We are able to see many things which are not primary light sources because light is scattered or reflected from them into our eye. For instance, when we look at a sheet of paper, light from a light bulb or from the Sun travels to the paper and then 'bounces' off the paper in all directions. This happens because the surface of the paper is full of microscopic irregularities, unlike the relatively smooth surface of a mirror. Some of the light enters the eye and we and others in the room see the paper. At the same time, some of the light falling on the paper may be absorbed. The key points here are that:

- light can be scattered or reflected from a surface;
- light can be partially absorbed by a surface;
- light travels from a primary source to an object and then to the eye so that we see it;
- nearly all solid material surfaces scatter or reflect light so most of the things we see are as a result of scattered or reflected light;
- solid materials which absorb most of the light falling on them are described as **black.**

Primary and secondary sources of light

When introducing the topic of light to primary school children, teachers should not assume that children have understood the difference between a **primary light source** and an object which scatters or reflects light (often called a **secondary light source**). In addition, children may not be ready to believe that light travels from one place to another and eventually arrives at the eye. With this in mind, it is often a good idea to begin a topic on light by focusing on things which give out their own light and how we might see them. The way we see light from a primary source is simpler than the way in which we see most things through scattered or reflected light. The light makes only a single journey from source to eye.

Make a collection of torches, candles and objects such as luminous glow-in-the-dark stickers and the chemical light sticks which emit light when broken. You can demonstrate how a toaster can glow safely when the lever is depressed. Try to arrange for children to see these turned on and off in a darkened room. If it is difficult to darken the room, then the children can use a torch and place objects in a cardboard box with a viewing hole cut in the side. When the eye covers the hole the box will be dark inside. Alternatively, use other safe primary sources of light under a thick blanket.

What happens to the torchlight when the light from it to the eye is blocked off? Cover the torch or cover the eye. Ask children to name more sources of light. Where does the energy come from each time?

RESEARCH SUMMARY RESEARCH SUMMARY RESEARCH SUMMARY **RESEARCH SUMMARY**

Warwick (2000) suggested that many children will not have experienced total darkness and will assume they could see in total darkness if they just try hard enough. Wadsworth (2000) described work on shadow puppets and how this can lead to a firmer understanding of light.

How does light travel?

Because light travels so quickly we find it difficult to appreciate that light travels at all. We know light can pass through a vacuum such as outer space because we can see the Sun and other stars, so, unlike sound, it does not require a medium to travel in. Scientists have found that light travels at about 300 000 kilometres per second (186 000 miles per second), which means that light from the Sun will take about 8 minutes to reach the Earth. Light from the next nearest star takes about 4 years to reach us so we are seeing that star as it was 4 years ago! Light travels slightly slower in air and even slower still in liquids such as water and solids such as glass.

For more information on energy, see Chapter 9.

If we return to the lighted candle we can think about how light travels from it. The energy associated with the candle wax is changed to light and heat energy when the candle is lit. In the same way, some of the energy associated with the Sun is changed into light energy, a small quantity of which reaches us here on planet Earth. Imagine how much of this light energy is lost to us as it spreads out into space! Other examples of light sources include television screens, electric lights, torches, a coal fire, an electric bar fire which glows red, a glow worm, the stars in the sky, fireworks, plankton in the sea (phosphorescence), an erupting volcano and a luminous watch face. All of these sources convert one kind of energy into light energy. In many instances the light energy is associated with fire or something burning.

The light from our candle now travels outwards in all directions. Just a little of this light goes in a straight line directly to the eye. The brain detects this as a very bright spot of light (the candle flame). The light which has escaped from the candle in other directions, however, will be reflected off the walls and floor and table and bounce around the room. Again, just a little of this ricocheting light will get to the eye and the brain will interpret this as seeing the things in the room.

Quite often, the way in which light travels from a source is shown as straight lines or light rays. This is both helpful and unhelpful. The light ray idea helps us to think of light travelling in a straight line from one place to another and perhaps reflecting from other objects until it reaches the eye. Sometimes, however, it is better to think of the light travelling outwards in a succession of spherical shapes rather like the layers of an onion until something gets in its way and blocks its journey.

Collecting evidence for light travelling in straight lines

We can tell children that light travels in straight lines but why should they believe us? More importantly, will they be willing to change their own ideas about how light travels if they do not already share the scientist's view?

Teachers can help children accept the scientific viewpoint by providing everyday examples of light travelling in straight lines.

- Children may have seen the light from a projector in a darkened room as it travels through smoke or dust – the light beam has a straight edge.
- Sometimes sunlight shines through clouds or tree tops when there is water vapour in the air – the Sun's rays can be seen to travel in straight lines.
- Shadows often have sharp edges – if light was able to travel around the object making the shadow (in a curve) then there would not be an area of darkness.
- A laser beam used at a disco or similar event can often be seen as a straight, pencil-thin beam.
- Children can look through a thin straight tube to see a light source at the other end – if the tube is slightly flexible the light will disappear when the tube is bent a little.
- A light source can be viewed through holes made in three cards, only if the holes are arranged in a straight line – a string pulled taut through the holes can be used to prove they are in a straight line.

How the eye receives light

Eyes allow humans and other living organisms to see. Human eyes are organs sensitive to light entering and falling on the **retina** at the back. Light energy affects the nerve endings in the retina which in turn produce tiny electrical impulses that are sent to the brain. The brain is important in that it interprets what the eye detects. For instance, certain nerve endings (the cones) on the retina in a healthy eye will detect only blue, red and green light. Combinations of these colours provide colour vision.

The brain can often be deceived by what the retina receives and that is why we prefer to use instruments to measure important phenomena. For example, when taking a photograph we often need to know, accurately, the intensity of light coming from a subject. The brain finds it difficult to judge this, so most cameras have a built-in light meter.

If we sliced a human eye vertically in half it would look something like Figure 11.1. Look at one of your own eyes and you will see the black spot called the **pupil** in the centre. While this is protected by the transparent cornea and aqueous humour, the pupil is, in fact, a hole into your eye which can change size. When too much light tries to enter the eye the iris will close the pupil slightly. If you walk into a darkened room, your pupils will enlarge to allow more light in. After a while you become accustomed to the dark, i.e. your brain is able to interpret what little light there is available entering through the enlarged pupil.

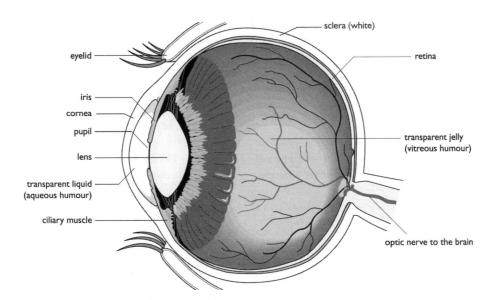

Figure 11.1 The human eye

If you punch a small hole in a sealed cardboard box, it will appear black because any light entering the box is unable to escape. (Remember, things which cannot reflect light are described as black.) This explains why a pupil appears black. When taking photographs of people we often get a 'red eye effect'. This happens because the light from the flash strikes the blood cells on the retina and is reflected directly back onto the film. Red eye reduction is achieved by a camera emitting a series of quick flashes before the photograph is taken. This forces the pupils to close down before the main flash.

RESEARCH SUMMARY RESEARCH SUMMARY RESEARCH SUMMARY **RESEARCH SUMMARY**

Many children believe that we see things by projecting something from our eyes (Harlen, 2000). In fact, the opposite is true and we see things only because light enters our eyes. The idea that rays are sent out from our eyes may be due to:

- comic strips depicting things like 'X-ray vision';
- expressions like 'looking daggers', 'staring' at someone or 'their eyes shone like pearls';
- the way we have to actively turn and focus on something in order to see it.

A strange experiment

The image of an object which forms on the retina is upside down. An experiment was carried out once where the subject was asked to wear a specially designed pair of glasses which turned the image the right way up. For a while the subject's brain interpreted the world as upside down. Soon, however, the brain adjusted and the subject was able to perceive the world correctly while still wearing the glasses. Once the glasses were removed, however, the subject had the misfortune to find that everything appeared upside down again! A while later the brain was able to readjust and link its interpretation once more to the physical world.

Light travels in waves

Light travels from a primary source, and is reflected from secondary sources, in waves which have particular characteristics. These waves are described as **transverse waves** and behave in a similar fashion to the waves on a pond when a stone is dropped in it. Pond waves consist of drops of water which move up and down while the wave shape moves at right angles to this, outwards.

Light waves and water waves have some important features that we can measure:

For more information on waves, see the section on sound waves in Chapter 12.

- the wavelength or the distance between one wave crest and the next;
- the number of waves passing in one second or the frequency of the waves;
- the amplitude of the wave or the maximum amount of up and down movement produced.

When thinking about light:

- the wavelength dictates the colour of the light – this might be measured in metres (e.g. red light has a wavelength of 0.0000007m);
- because all light waves travel at the same speed the shorter waves have a higher frequency – measured in kilohertz (kHz);
- the amplitude of the wave dictates the intensity or brightness of the light. More energy is required to produce larger water waves and in the same way a more intense light is associated with greater energy.

The electromagnetic spectrum

Light waves are part of a large family of waves called the **electromagnetic spectrum**. We are already familiar with many of the types of wave in this spectrum. Ultraviolet radiation, X-rays and gamma rays have wavelengths shorter than visible light. Infrared radiation, microwaves and radio waves have wavelengths longer than visible light. Table 11.1 shows the tiny section of the electromagnetic spectrum in which visible light waves can be found. These are generally only about one-thousandth of a millimetre in length. Red light is to the lower end of the light spectrum with relatively long waves and violet is at the other end. This shows how the colour of the light is related to its wavelength.

Approximate wave length	Type of electromagnetic wave	Notes
1/100 000 000 of a millimetre	gamma rays	High energy and dangerous
1 millionth of a millimetre	X-rays	'Shadows' produced by X-rays used to photograph bones, etc. in hospitals
1/10 000 of a millimetre	ultraviolet	Harmful to the skin. Many of these rays are blocked by the Earth's atmosphere and sun creams
1/1000 of a millimetre	visible light – violet, indigo, blue, green, yellow, orange, red	The visible spectrum. Violet has the shortest wavelength, red has the longest
1/100 of a millimetre	infrared	These carry the heat from an electric fire. Infrared can be 'seen' with special night-vision binoculars and heat-seeking cameras
10 millimetres (1 centimetre)	microwaves	Used in microwave ovens
1000 millimetres (1 metre)	ultra high frequency (UHF)	Television broadcasts
1000–10 000 metres	Very high frequency (VHF) and other radio waves	Include FM and long-wave broadcasts

Table 11.1 The electromagnetic spectrum

Making shadows

When light waves are blocked by something, a dark area is formed called a **shadow**. A shadow is the absence of some of the available light. Often light arrives at the shaded area from other directions, so a shadow is rarely completely **black**.

Shadows can have sharp edges or fuzzy edges. When there is a single light source which is very small, then sharp-edged shadows can be made. A good example of this is when the Sun is shining. When the light source is spread over an area or there is more than one source, then a fuzzy-edged shadow will be formed which has an umbra and penumbra (see Figure 11.2). Shadows will form on the side of an object which is away from the light source. In a well-lit office or classroom, there will be very few sharp shadows because the number of electric lights ensures that most areas are well illuminated from a variety of directions.

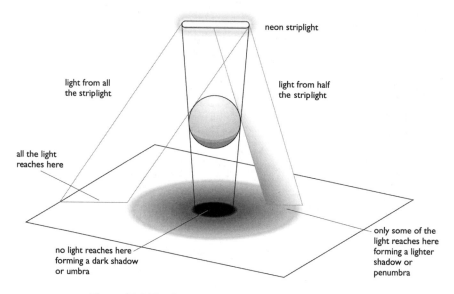

Figure 11.2 The fuzzy-edged shadows made by a ball

PRACTICAL TASK PRACTICAL TASK PRACTICAL TASK PRACTICAL TASK

Light is one of the richer sources of investigative activities. EYFS children can learn a great deal from trying to jump on their own shadow on a sunny day. Playing with the shadows made by PE hoops involves asking 'What happens if …?' and 'What happens when …?' Key Stage 1 children can experiment with the different sizes of shadows made by moving the light source closer to or further from the object casting the shadow. At Key Stage 2, children should be trying to change the factors which alter the size of shadows. The resulting graphs challenge the abler children to think about infinitely large shadows.

RESEARCH SUMMARY RESEARCH SUMMARY RESEARCH SUMMARY RESEARCH SUMMARY

Children often mix up the words 'shadow' and 'reflection'. They might say, for example, 'I think a shadow is a reflection from the Sun'. Harlen (2000) suggests that children need a great deal of practical experience of shadows before they are really ready to talk about them as a phenomenon caused by the blocking of light.

More about reflected light

When light hits the surface of an object some of it may be absorbed causing the surface to warm up slightly. Some light will also be scattered or reflected even by a matt surface. This means that the tiny irregularities in the surface will make the light rays bounce off in all directions.

When light travels to a polished surface, such as a mirror, most of the light will be **reflected** in a regular fashion and not scattered. A beam of light will bounce off the mirror just as a ball bounces off a smooth wall. The angle between the approaching ray and the mirror and the reflected ray and the mirror are the same (see Figure 11.3).

A mirror gives the impression that the object is behind it because the brain expects light from objects to travel in straight lines. Figure 11.3 shows how the image in a mirror seems to be the same distance behind it as the object is in front. Once the surface of the mirror becomes curved this begins to play tricks with where the image appears to be and this gives rise to distorted images.

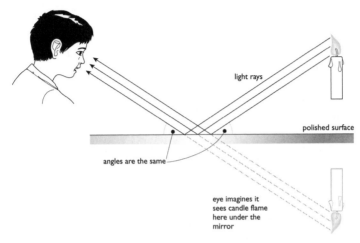

light rays

polished surface

angles are the same

eye imagines it
sees candle flame
here under the
mirror

Figure 11.3 Reflections in a mirror

PRACTICAL TASK PRACTICAL TASK **PRACTICAL TASK** PRACTICAL TASK

At Key Stage 1, encourage the children to play with plastic mirrors. Can they see round a corner? Can they look over a desk? Can they use two mirrors to see the back of their head? At Key Stage 2, get two small mirrors and tape them together along an edge. Alternatively, hold one small mirror up to a larger one so that they touch. Place an object such as a colourful pen top between the two mirrors near the 'fold'.

Experiment to find out how many images you see for different angles between the mirrors. Make a table of your results and try to spot any patterns that emerge. Work out which angles you should hold the mirrors apart in order to get complete numbers of images. How many images do you see when the angle is 180 degrees?

Use this experience to plan a simple lesson involving mirrors and the skills of scientific enquiry.

Coloured light

The nerve cells which respond to colour on the retina of a human eye can detect only blue, red and green light – the primary colours of light. When all three colours fall on the retina in the correct proportion, the brain interprets this as white light.

When different combinations and proportions of these three colours fall on the retina, we see these as the wide range of colours that we are familiar with.

We see the colour of objects because that colour is mostly scattered or reflected from the surface of the object, all other colours being absorbed. For instance, when daylight is reflected from a blue ball, mostly blue and some green light is scattered from its surface. Most of the red light is absorbed by the surface and never reaches the eye. The brain interprets this as seeing a blue ball. Very few colours that are scattered or reflected from objects around us are pure blue, red or green so it is useful to think in terms of *mostly* blue light reaching the eye with some green (and even a little red), which can be interpreted as the particular blue of the ball we see.

Similarly when we see a yellow daffodil it is because the flower reflects mostly red and green light onto the retina and the brain interprets this as yellow. The leaves of the daffodil absorb most colours in order to make food for the plant through photosynthesis. Green light, however, is generally not used in this process so most plant leaves scatter green light.

Mixing coloured light

When coloured light from two or more different sources falls on a surface, the intensity of the light scattered from that surface increases. Blue, red and green, the primary colours of light, combine to give white. Magenta, cyan and yellow are secondary colours formed by mixing two primaries in equal proportions:

- blue + red light = MAGENTA – a vivid pinkish red;
- blue + green light = CYAN – a peacock blue;
- green + red light = YELLOW;
- blue + red + green light = WHITE.

Coloured pigments

The artist has a set of primary colours which are not the same as the primary colours of light. These are red, blue and yellow. Red pigment (paint for instance) scatters mostly red light. Yellow pigment, however, scatters mostly red and green light, absorbing blue light. Blue pigment scatters mostly blue and some green light, absorbing red light. Mixing yellow and blue pigment therefore allows only green light to be scattered, all other colours being absorbed (see Figure 11.4).

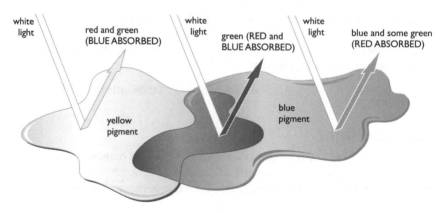

Figure 11.4 Mixing pigments

Notice that the mixed green pigment reflects less light than the two other pigments. The green is of a lower tone than its constituent colours. Mixing coloured pigments:

- blue + yellow pigment = GREEN;
- yellow + red pigment = ORANGE;
- blue + red pigment = PURPLE;
- blue + red + yellow pigment = BLACK (usually dark brown).

Green, orange and purple are also referred to as secondary colours. Things which are described as black will absorb most of the light falling on them. This means that they will get a little hotter as a result. Dark clothing, worn on a hot sunny day, will make us feel hotter, while lighter coloured clothing will reflect much of the electro-magnetic spectrum, enabling us to feel cooler.

Colour filters

Coloured sweet wrappers that you can see through and stained glass windows are examples of **colour filters**. A blue sheet of cellophane is a blue colour filter which allows only the blue part of white light through (see Figure 11.5). A filter is a bit like a colander. When you strain peas in a colander, the water passes through and the peas remain blocked. A magenta colour filter allows red and blue light through (the water) and stops the green light (the peas).

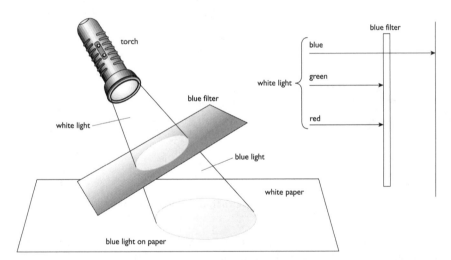

Figure 11.5 Colour filters

PRACTICAL TASK PRACTICAL TASK **PRACTICAL TASK** PRACTICAL TASK

Give EYFS and Key Stage 1 children torches with coloured filters or coloured sweet cellophane over the beam. Work in a darkened area and make coloured shadows. Use two different colours over two torches for magical colour-effect shadows.

With Key Stage 2 children, use a strong light source such as an overhead projector and experiment by making shadows on a screen or the wall with different materials which are transparent, translucent or opaque. Try using some colour filters to make shadows.

Light passing through transparent material

Light is able to pass through materials which are transparent or translucent. Some light is always absorbed and, in the case of a translucent material such as frosted glass, the light rays may be scattered so that no clear image is seen.

When light rays appear to change direction while passing from one transparent medium to another this is called refraction. Refraction enables lenses to work and gives rise to the impression that water is not as deep as it really is. Light passing from air into a glass prism is refracted so that it is split into its component colours. This dispersion of light occurs because the speed at which light travels is different in different materials.

Waves can help us to understand this. Waves on the sea are formed by the action of the wind and generally follow the wind direction. As they approach land, however, parts of each wave will be slowed down, causing the wave to change direction. A glass prism has a similar effect on the waves of light entering it (see Figure 11.6). Different colours will be refracted at different angles (because they travel at different speeds) and so the beam coming out of the prism will be split into the colours of the spectrum.

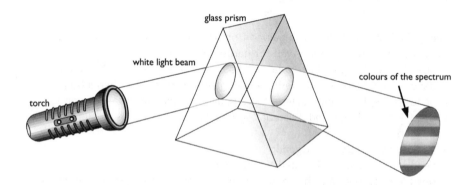

Figure 11.6 Light is refracted through a prism

A rainbow is formed when sunlight is refracted and reflected a number of times inside each spherical drop of rain. The refraction enables the white light to be split up into the colours of the spectrum while the reflection means that, with our back to the Sun, we receive the light from the rainbow as though it were a mirror in the sky.

A SUMMARY OF **KEY POINTS**

> **Light is a form of energy.**

> **Light travels in a straight line in waves from a source.**

> **Light can be scattered or reflected from an object.**

> **We see because light from a source or scattered from an object enters the eye – the brain interprets the effects of light on the retina.**

> Light is just a small part of the electromagnetic spectrum of waves, which also includes radio, television and X-rays.

> Shadows are made by the absence of light which may have been blocked.

> Light can be reflected from a smooth surface such as a mirror just as a ball bounces from a smooth wall.

> The colour of light is dependent on its wavelength.

> When blue, red and green light strike the retina the brain interprets this as white light.

> White light can be separated into the colours of the spectrum – violet, indigo, blue, green, yellow, orange and red.

> Coloured objects reflect mostly light of that colour and absorb the other colours.

> A coloured filter allows one or more colours through and absorbs the rest.

> Light passing through transparent or translucent materials may have its direction changed.

M-LEVEL EXTENSION > > > > M-LEVEL EXTENSION > > > >

Look back over the practical activities suggested in the chapter, including the research summaries, for helping children to understand light. Divide them up into those that could take place in the classroom or other indoor space and those that would be best undertaken outside. Focus on how you might use the outdoor classroom and playground areas to best effect in your planning of this aspect.

REFERENCES REFERENCES REFERENCES REFERENCES REFERENCES

Harlen, W. (2000) There's more to light than meets the eye. *Primary Science Review*, 64, 20–22.

Wadsworth, P. (2000) Festivals of light. *Primary Science Review*, 64, 28–30.

Warwick, P. (2000) Light and dark. *Primary Science Review*, 64, 23–25.

FURTHER READING FURTHER READING FURTHER READING FURTHER READING

DfE (2011) *Teachers' Standards*. Available at www.education.gov.uk/publications.

Howe, A., Davies, D., McMahon, K., Towler, L. and Scott, T. (2005) *Science 5–11: A Guide for Teachers*. London: David Fulton.

Wenham, M. (2005) *Understanding Primary Science – Ideas, Concepts and Explanations*. London: Paul Chapman Publishing. This contains a useful chapter on all aspects of light.

12
Sound

Curriculum context

National Curriculum programmes of study

At Key Stage 1, children should be taught that there are many kinds of sound and sources of sound, that sounds travel away from sources, getting fainter as they do so, and that sounds are heard when they enter the ear.

At Key Stage 2, children should be taught that sounds are made when objects vibrate but that vibrations are not always directly visible, how to change the pitch and loudness of sounds produced by some vibrating objects, and that vibrations from sound sources require a medium through which to travel to the ear.

Early Years Foundation Stage

Children must be supported in developing the knowledge, skills and understanding that help them to make sense of the world. By the end of the EYFS, children should:

- investigate objects and materials by using all of their senses as appropriate;
- find out about, and identify, some features of events they observe;
- look closely at similarities, differences, patterns and change;
- ask questions about why things happen and how things work.

Introduction

Most children already know a lot about sound. They will tell you that sounds can be of different loudness and can be pleasant or otherwise to listen to. Many children will tell you that sounds are made by vibrations but may not really know what a vibration is. Some children will understand that sounds have pitch and quality without necessarily understanding why this varies.

The concept of sound is relatively easy to understand because we can use the senses of hearing, touch and sight to explore its many aspects. Sound is a form of energy, a physical thing which depends on objects vibrating in certain ways, and this movement can be felt and observed as well as heard.

What is sound?

- Were there sounds on this planet billions of years ago when there were no ears to hear them?
- Are there sounds on the surface of the Sun?
- Does a person suffering from tinnitus (an unwelcome ringing in the ears) really hear sounds?

The answers to these questions depend on our definition of **sound**. It is helpful to think of sound as the result of vibration together with the effect this has on the ear and the brain. This is really two definitions in one, involving the source of the sound as well as the receiver. Sound needs something to travel in such as water, air or solids. It cannot travel across the vacuum of outer space, so we do not hear the result of explosions on the surface of the Sun, and astronauts within a few centimetres of each other can talk to each other only by radio.

Sounds come from a vibrating source

Vibrations happen when something moves backwards and forwards (or up and down, etc.) in a regular pattern. Think of a ruler held tightly to a desk. If you 'ping' it, you can see it moving up and down rapidly in a regular way – each up and down movement takes about the same time. A plucked guitar string can be seen to move back and forth rapidly forming a smooth arc as it does so.

PRACTICAL TASK PRACTICAL TASK **PRACTICAL TASK** PRACTICAL TASK

You can demonstrate for yourself how vibrations behave in slow motion. Set up a simple pendulum by hanging any object from a length of string. Set the pendulum swinging and watch what happens.

- Use a stopwatch to time 10 complete swings, then time the next 10 complete swings. You should notice that the time for each swing is the same at the beginning of the activity as towards the end.
- What happens to the distance the pendulum bob moves as time goes by?
- When is the bob moving the fastest?
- When does the bob stop moving for a fraction of a second?
- Think of other things which behave like the pendulum such as a rocking chair or a metronome.
- At Key Stage 1, children can compare the swing of a long pendulum with that of a short one.

A child on a swing shows how vibrations behave, but in slow motion. Each to and fro movement takes about the same time and, if the child sits still, the swing gradually comes to a halt. If the child adds energy to the swing (by swinging their legs, for instance) then the movement will continue. We know that vibrations become gradually smaller and smaller unless energy is provided to keep them going. The note from a piano will die away while the electrical energy supplied to a buzzer will keep it buzzing.

Giving children experience of vibrations

Children need to build up their concept of vibrations through a range of practical activities.

- Ask them to feel their voice box at either side of their throat with their fingers as they speak. Can they feel the greater intensity of vibrations when they speak louder?
- Children can see the vibrations when they 'ping' an elastic band or a ruler held on a desk.
- A tuning fork which has been set vibrating will disturb the surface of water if it is gently lowered on to it. The vibrating fork will also make a loud buzzing sound when held to a sheet of paper. The fork can also be made to 'kick' a suspended table tennis ball.
- Feel the vibrations in a plucked guitar string.
- See grains of rice jump around on the surface of a drum when it is hit gently.
- See grains of salt jump around on a glockenspiel bar as it is gently tapped.
- See and feel the vibrations of an exposed loudspeaker cone – especially at low frequencies.
- See slow-motion video clips of vibrating objects.

Sounds travel outwards from a source in waves

A sound made by a human voice will travel across a room and be heard on the other side. The initial sound will travel in all directions from the source in shapes of ever-expanding spheres a bit like the layers of an onion. These spheres of movement expand in waves which have properties in common with other waves. As the waves move away from the source their energy is dissipated over an ever-increasing area so the sounds become weaker at any one point. Thus, a listener will hear weaker sounds as the source moves further away.

Sound waves and air molecules

Sound waves are created in air because the vibrating source disturbs the air around it. The to and fro movement of the vibrations will alternately push the air molecules closer together (called **compression**) and then pull the air molecules apart (called **rarefaction**). Think of the skin of a drum when it is struck for the first time – it will press the air molecules nearest to it closer together (see Figure 12.1). This area of compressed air will, in turn, push on the set of molecules next to it and so on. The area of compressed air moves outwards in a regular wave.

Imagine the child sitting on the swing is joined by others on a row of swings. If the first child begins swinging sideways they will begin to bump into the swing next to them. The second swing will begin to sway back and forth and bump into the next swing and so on. In this way a wave of swinging can travel from one end to the other in a similar way to the way sound waves travel. Notice that the children themselves do not move along, just the swinging effect. In the same way, the molecules of air do not move very far (only back and forth) but the wave of vibra-

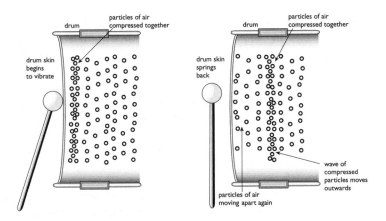

Figure 12.1 How the molecules of air behave as they vibrate

tions does travel. Thus, the air does not flow from the source to the ear – an idea often held by children.

Sound waves are not the same as light waves nor the waves on a pond of water. On water the particles of water bob up and down while the wave moves outwards horizontally. With sound, the particles of the substance move back and forth in the *same direction as the wave is travelling*. This is an example of a longitudinal wave.

For more information on light waves, see Chapter 11.

Using a 'slinky'

A very good way of demonstrating how the molecules of a substance behave when transmitting a sound is to use a 'slinky' spring extended on a smooth surface such as a desk (see Figure 12.2). Push one end rapidly and a wave is sent along the length of the spring. It is possible to see the wave bounce back (an echo) and of course to see that the coils of the spring (representing the molecules) do not move along with the wave but merely 'vibrate' back and forth.

Sound waves behave like all other waves in that they can be reflected (to make an echo), refracted and absorbed by substances.

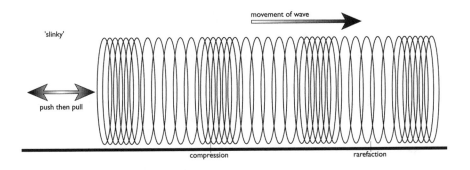

Figure 12.2 Using a 'slinky' to demonstrate sound waves

For more information on solids, see the section on solid, liquid and gas in Chapter 6.

Sound travels in a medium

Sound will travel in solids, liquids and gases such as air. Our experience of sound travelling in solids and liquids tends to be limited and we often think of it as being less efficient than travelling in air. For instance, if we close a solid door it tends to prevent sound escaping from a room. This, however, is because much of the sound will have been reflected by the door back into the room or absorbed by the material of the door.

Sound actually travels faster and more efficiently through solids than liquids or gases. This is because of the arrangement of the molecules in a solid which makes it possible to transmit the sound wave more readily. By the same argument, sound travels better in liquids than in gases. There are various ways in which we can help children understand that sound travels better in a solid or liquid than in air.

RESEARCH SUMMARY RESEARCH SUMMARY **RESEARCH SUMMARY RESEARCH SUMMARY**

Harlen (2007) highlights research into children's knowledge and understanding of sound, showing a progression in the way that they perceive the causes of sound. In the early stages of understanding, children will associate sounds with a physical action but not be aware of vibrations being set up ('I think the drum makes a sound because it is hollow and it echoes'). Later, they will observe the vibrations but may mix up cause and effect ('the sound made the vibrations'). A few children at Key Stage 2 will make the correct association between vibrations and sound.

Children who understand that they need ears to hear sounds may not necessarily be aware of their internal workings. They may not understand that what reaches the brain are tiny electrical impulses which in turn must be interpreted by the brain to register as a type of sound.

Perhaps the most difficult concept to understand is that sound travels in waves through a medium. Children at Key Stage 2 will often think that the air carries the sound by moving with it ('sound is a special piece of air going into our ears') and will have little idea that sound travels more efficiently in liquids and solids.

Chang (2007) conducted investigations into the concepts held by Taiwanese children about the production and movement of sound. He found that over half of junior age pupils represented sound as a wave. However, one quarter of the children thought that if a sound-producing machine was sealed inside a container, then no sound would be heard. He cites Vienott (2001) who reports that children do not take into account the medium through which the sound travels. Many children assert that sound just travels. Chang refers to earlier studies that show that children associate sound transmission with gaps and holes rather than vibrations.

Shipman (2006) uses models with children acting as particles to demonstrate how sound travels as waves. He shows how modelling of this kind can lead to more abstract thinking about the properties of waves. These ideas include pitch, frequency, loudness, energy, echoes and vibrations. Shipman is careful to point out the limitations of any model, but he indicates useful animations which show sound. Watkins and Shepherd (2006) use discussion with audiologists to help illustrate many of the features of sound we teach at primary level.

Solids

- Children put their ear to a desk top and scratch the desk very gently. Can they hear the same scratching sound through the air when their head is lifted?
- If a protected ear is placed on a central heating radiator, then the sound of the pump can often be heard or alternatively someone tapping another radiator elsewhere in the building.

- Children may have heard stories of people who can hear horses or a train approaching by putting an ear to the ground.
- Try putting a drinking glass to a wall and listening to noises on the other side.
- Make a string telephone in which the sound waves are carried by a taut string between two tins.
- Put an ear to the school railings while someone some distance away taps them.

Liquids
- Place a waterproof wristwatch in a jar of water and listen to the alarm going off.
- In the swimming pool make sounds under the water. They can be heard clearly by someone swimming under water.
- Children can read about how some sea creatures such as dolphins and whales communicate over long distances by sounds.

Receiving vibrations at the ear drum

In humans, vibrations from a sound source will travel through the air and be funnelled by the outer ear (the pinna and ear canal) towards the ear drum. As the drum is made to vibrate in unison with the air molecules, it moves some tiny bones in the protected middle ear. These in turn stimulate the liquid in the cochlea which is situated in the inner ear. There are tiny hairs in the liquid inside the cochlea which respond to different vibrations. It is these hairs which enable electrical impulses to be sent to the brain which then registers the sound (see Figure 12.3).

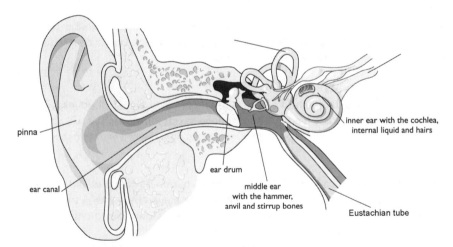

pinna

ear canal

ear drum

middle ear
with the hammer,
anvil and stirrup bones

inner ear with the cochlea,
internal liquid and hairs

Eustachian tube

Figure 12.3 The ear

The properties of sounds

Sounds have a number of simple properties which can be observed by children and can be understood in terms of the kinds of vibrations made. Sounds can:

- vary in loudness;
- vary in pitch;
- vary in quality.

The loudness of sounds

If a stretched elastic band is plucked it can be made to make a quiet sound by 'pinging' it gently. The amount the band moves from the centre line is small. To make a louder sound you would pull the band further to one side and let it go. The amount that the band moves from the centre line is called the amplitude of the vibration. The greater the **amplitude** the louder the sound (see Figure 12.4). **Loudness** is measured in decibels. Hitting a drum skin hard will produce a loud sound because more energy is put in and the skin is made to move further from its starting point. As the sound made by a plucked guitar string dies away it becomes quieter because the amplitude of the vibration is getting smaller.

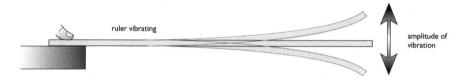

Figure 12.4 The amplitude of a vibrating ruler

Sounds can be amplified or made louder in a number of different ways.

- By using more energy to make the sound. A drum hit harder will make a louder sound.
- By using electronics. A hi-fi amplifier uses electronic circuitry and electrical energy to create louder sounds.
- By using a resonating box. The violin, guitar and drum all use the principle of a hollow box of the correct dimensions to make sounds louder. This can also be observed in a shower cubicle.
- By using a funnel to direct the sound in a particular direction.

The pitch of sounds

In children's minds the **pitch** of a note is often confused with its loudness. This is probably because we use a confusing language to describe both. We will often talk about turning up the volume to a higher level or using the word *high* to describe the pitch of a note.

If a whistle represents a high-pitched note then a humming sound would represent a low pitch. On a piano the high-pitched notes are played by the keys towards the right-hand end. A double bass will tend to play mostly low-pitched notes.

The pitch of a note is related to the number of vibrations made per second or the frequency of vibrations. Frequency is measured in hertz (Hz) which is one vibration per second or kilohertz (kHz). The note we call middle C on a piano is produced by a **frequency** of 256 Hz. The higher the frequency, the higher the pitch of the note, so the note one octave above middle C is caused by vibrations of 512 Hz.

There are a number of factors that affect the pitch of a note which can easily be investigated by children. The pitch of a note is affected by:

- the length of a vibrating string or column of air. Think of shortening a guitar string by holding it down at a fret. This produces a higher-pitched note. The vibrating air in a short organ pipe will produce a relatively high-pitched note;

- the relative size of a vibrating object. Think of the bars on a glockenspiel or xylophone or a tuning fork. The smaller ones give a higher-pitched note;
- the tension in the object making the note. If you support an elastic band over an open box and pluck it, a higher note will be achieved when the band is stretched further. A drum skin will give a higher note when it is stretched further, and a stringed instrument can be tuned to a higher pitch by turning the tensioning screw.

The human ear can detect some frequencies but not others. If an object vibrates with a frequency greater than about 20 Hz then this will produce a sensation of hearing in humans. Furthermore, if the object vibrates faster than about 20 000 Hz, then a human will probably not hear anything. Some animals have a greater frequency range than humans, so a dog, for instance, will respond to a high-pitched whistle that a human cannot hear.

PRACTICAL TASK PRACTICAL TASK PRACTICAL TASK PRACTICAL TASK

Get hold of two identical glass bottles such as milk bottles. Half fill one with water and fill the other three-quarters full.

- Tap each bottle and decide which gives the highest pitch note. (You may find it useful to try to sing each note and then decide which is the highest pitch.) When you tap the bottles the combination of glass and water vibrate to make the sound. The high-pitched sound should be achieved with the least water. This reinforces the idea that less material will generally produce a higher-pitched sound. (Think of the glockenspiel bars.)

- Blow horizontally across the top of each bottle to make a musical note. Use the information gained above to explain how you got the high-pitched note this time. Which substance was vibrating this time? The high-pitched note is formed by the shorter column of air which vibrates.

The quality of sounds

Sounds can vary in **quality**. A tuning fork will produce a relatively pure musical note which would be displayed as a simple wave on an oscilloscope. Some musical notes consist of a basic sound wave (the fundamental) together with frequencies which are multiples of this called harmonics or overtones. Most other sounds and especially noises are made up of a collection of different, unrelated waves. Their display on an oscilloscope would be a fairly wild, wavy line which was changing in shape all the time. Furthermore, a musical note made by an instrument such as a trumpet will not sound the same as the same note made by a piano or the human voice. Each instrument produces a collection of additional vibrations to accompany its fundamental note. Unwanted or unpleasant sounds are often referred to as noise.

The speed of sound in solids, liquids and gases

Sounds can travel at approximately 6 000 metres per second in some solids and at a quarter of this speed in water. Sound can travel in air at approximately 332 metres per second. This is fast, but not nearly as fast as light which travels at 300 000 *kilometres* per second.

For more information on light travel, see Chapter 11.

This difference in speeds enables us to appreciate that sound does take time to travel. When we see lightning the sound it produces *at exactly the same time* is often heard as thunder a few seconds later by an observer a few miles away.

Unfortunately children will not always accept that the light and sound are from the same instantaneous release of energy. It is, however, possible to watch a sound being made at some distance away and to detect a slight delay in hearing the sound. Exploding fireworks, the click of a ball on a cricket bat and a child bashing a dustbin lid at the other end of a playing field will all provide this opportunity.

The loudspeaker and the microphone

For more information on energy, see Chapter 9.

Sound energy can be changed into electrical energy and vice versa. A simple microphone will detect the tiny movements caused by sound waves and convert these into electrical impulses. These impulses are often amplified before being transmitted as radio waves or converted back into sound in a loudspeaker. The telephone enables sound to be converted into electrical impulses which are trans-mitted over great distances by radio waves, fibre optic cables or conventional electrical wires. The same telephone receiver will receive small electrical impulses which are converted back into sound energy in the earpiece.

For more information on electricity, see the section on electrical charge in Chapter 8.

The simple idea which enables this to happen is that when a coil of wire is moved near to a strong magnet, small electrical impulses are set up. If the coil of wire is made to vibrate then the impulses will be produced that match the vibrations.

The reverse is also true. Small electrical impulses flowing through a coil of wire will make it move in the presence of a strong magnet. If the coil is attached to a cone of card or similar material this will set up sound vibrations in the air. This idea can be explored by children if they wrap a coil of wire around a magnetic compass and connect the wire to a battery. The magnetic needle will move when this happens. Children might like to look inside an old, disconnected loudspeaker to see the coil and card cone.

The oscilloscope

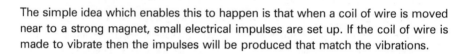

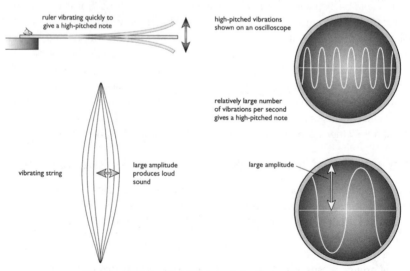

Figure 12.5 Typical displays on an oscilloscope

An **oscilloscope** is used to convert sound into a display on a screen. The display will give a graphic picture of the sound waves, which are informative but can be misleading. The waves displayed resemble **transverse waves** like those produced on water but, of course, sound waves are not like these. However, the display *will* show the increased frequency of the waves as the pitch of the sound rises. The amplitude of the waves can also be shown to increase as the loudness is increased (see Figure 12.5).

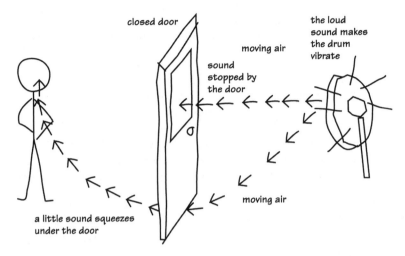

Figure 12.6 Child's drawing of how sound is heard through a door

A SUMMARY OF **KEY POINTS**

> Sound occurs when an object vibrates at a suitable rate and these vibrations are detected by the ear and interpreted by the brain.

> Sound needs a medium to travel through, which can be a solid, liquid or gas.

> Sound will not travel in a vacuum.

> Sounds leave a vibrating source in all directions, but their energy becomes dissipated as they move further from the source.

> Sounds travel in longitudinal waves which can be reflected, refracted and absorbed.

> Sound waves have amplitude which affects the loudness of the sound.

> Sounds can be amplified in a variety of ways.

> Sound waves have frequency which affects the pitch of the sound produced.

> Sound can be converted into electrical impulses in a microphone.

> Sound can be made by converting electrical impulses into physical vibrations in a loudspeaker.

> The quality of a sound depends on the combination of sound waves that produce it.

M-LEVEL EXTENSION > > > > M-LEVEL EXTENSION > > > >

Understanding sound is another aspect of science that may benefit from setting in a cross-curricular context. An obvious link can be made to music, exploring the sounds made by a range of different instruments, including tuned and untuned percussion, and discussing tone, pitch and timbre. But there are other curriculum areas and subjects that you could consider, for example designing sound-making machines and finding out how microphones, loudspeakers and radios work in technology, and using ICT programs to add sound effects to a story or play that the children have written and recorded in their literacy work. Think about how this could be planned for the particular yeargroup you are working with or across each of the primary phases.

REFERENCES REFERENCES **REFERENCES** REFERENCES REFERENCES

Chang, H. (2007) Investigating primary and secondary students' learning of physics concepts. *Taiwan International Journal of Science Education*, 29(4), 465–82.

Harlen, W. (2007) The SPACE legacy. *Primary Science Review*, 97, 13–16.

Shipman, B. (2006) Just how does a sound wave? *Primary Science Review*, 93, 12–14.

Vienott, E. (2001) *Reasoning in physics: The part of common sense*. London: Kluwer.

Watkins, R. and Shepherd, K. (2006) Are we all listening? *Primary Science Review*, 93, 15–17.

FURTHER READING FURTHER READING **FURTHER READING** FURTHER READING

DfE (2011) *Teachers' Standards*. Available at www.education.gov.uk/publications.

Howe, A., Davies, D., McMahon, K., Towler, L. and Scott, T. (2005) *Science 5–11: A Guide for Teachers*. London: David Fulton.

Wenham, M. (2005) *Understanding Primary Science*. London: Paul Chapman Publishing.

13
The Earth and beyond

Curriculum context

National Curriculum programmes of study

There is no longer a statutory requirement to teach the Earth and beyond at Key Stage 1, though some experienced teachers continue to do so. Teaching about 'time' (e.g. hours in a day, days of the week, months of the year) and 'change' (e.g. sequencing daily events, seasonal variations in the local environment) does provide a valuable introduction.

At Key Stage 2, children should be taught about the Earth, the Sun and the Moon and periodic changes associated with the Earth–Sun–Moon System. Aspects of the Solar System and wider Universe, including the history of space exploration, should be drawn from the Key Stage 3 programme of study (statutory requirements allow for this to be done). The Solar System and wider Universe significantly enhance an otherwise 'dry' minimum entitlement and provide a more balanced, stimulating and appropriate body of content.

Early Years Foundation Stage

Children must be supported in developing the knowledge, skills and understanding that help them to make sense of the world. By the end of the EYFS, children should:

- find out about, and identify, some features of events they observe;

- look closely at similarities, differences, patterns and change;
- ask questions about why things happen and how things work;
- observe, find out about and identify features in the place they live and the natural world.

Introduction

It is often said that many of the more 'abstract' ideas associated with the Earth and beyond and space in general are far too difficult for children in the primary years and that in any case it would be inappropriate to teach them in terms of the 'hands-on' way in which primary science 'is done'. Evidence from classroom-based research studies and changes in the ways in which the 'doing' of primary science is perceived have demonstrated that such views are largely unfounded. The exploration of space is challenging, exciting and very much a part of our everyday lives. Children of all ages find this area of science particularly fascinating and motivating.

RESEARCH SUMMARY RESEARCH SUMMARY RESEARCH SUMMARY RESEARCH SUMMARY

Kibble (2002) pointed out that there is a considerable body of research material covering children's ideas about Earth in space. Many young children see themselves at the centre of the universe, with the stars, sun and moon in the sky above them. Kibble used a set of cards to elicit and challenge ideas held by the children about day and night, gravity, seasons and phases of the moon. Much of it was lost or reshuffled into Key Stage 3 to make room in a science curriculum which, from the outset, was seriously overcrowded. The history of astronomy in the National Curriculum for Science was reviewed by Sharp and Grace (2004).

The Universe, galaxies and stars

The Universe

The **Universe** is, quite literally, everything that exists.

For more information on atoms and molecules, see the section on what materials are made of in Chapter 6.

- Matter (from atoms and molecules to stars and galaxies).
- Radiation (visible light together with the rest of the electromagnetic spectrum).
- Space (the vast 'emptiness' both within and between galaxies).

The Universe is about 14 billion years old. It most probably emerged from an explosive event generally referred to as the **Big Bang**. Fourteen billion years ago, the Universe looked very different from how it looks today. Then, it was a small, unimaginably hot, unimaginably dense primordial fireball filled only with radiation. It took until about 1 million years after the Big Bang for the Universe to have expanded and cooled sufficiently for the first clouds of hydrogen and helium gas to appear, and about another 1 billion years after that for some of the hydrogen and helium to come together under the influence of gravity to form the first **galaxies**. Hydrogen and helium are the simplest and remain the most abundant elements in the Universe as a whole. The Universe has been expanding and the distances between galaxies increasing ever since.

Whether the Universe will continue to expand or not depends on the amount of matter contained within it. If sufficient, gravity will slow down its rate of expansion and pull the whole Universe back together again. Galaxies will inevitably collide and everything will end in a Big Crunch. If not, the Universe will simply get bigger and bigger. The stars within galaxies will eventually burn out and die and everything will end in a Big Freeze.

Galaxies

The Universe contains countless galaxies. Galaxies are assemblages of **stars**, **nebulae** and other interstellar material. A typical galaxy contains about 100 billion stars and measures about 100 000 light years across. A light year is the distance that light travels through the relative vacuum of space in one year (about 10 million million kilometres). Galaxies are classified into four main groups depending on their appearance.

1. Spirals (about 20 per cent of all known galaxies, disk-shaped with bulbous central regions or nuclei and bright spiral arms).

2. Barred spirals (about 10 per cent of all known galaxies, similar to spirals but with central regions consisting of stars arranged in the shape of a bar).

3. Ellipticals (about 60 per cent of all known galaxies, generally smaller than spirals and barred spirals, with no arms).

4. Irregulars (about 10 per cent of all known galaxies, generally smaller than all other galaxies, with no defined shape).

Galaxies are not scattered randomly throughout the Universe, they occur in clusters.

- Rich clusters (may contain hundreds or thousands of galaxies).
- Poor clusters (may contain as few as ten).

Clusters of galaxies may themselves form superclusters which spread over regions of space up to 100 million light years across. Superclusters are the largest known structures of the Universe.

Our own Sun, our nearest star, is located within a galaxy referred to as *the Galaxy* or the Milky Way. The Milky Way is also a term used to describe the faint band of stars that can be seen running across the sky on a clear night (the plane of *the Galaxy* viewed from within). The Milky Way is possibly a spiral galaxy with a disk about 80 000 light years across. The Milky Way formed about 10 billion years ago, some time after the Big Bang. The Sun is located within the Orion Arm of the Milky Way, about 25 000 light years from the central bulge or galactic nucleus. It has been estimated that as the Milky Way rotates it takes the Sun about 200 million years, travelling at about 230 kilometres per second, to orbit the galactic nucleus once. The Milky Way belongs to a poor cluster of about 30 other galaxies known as the Local Group.

Stars

Typical stars are balls or spheres of hot glowing gas which form within the nebulae of galaxies when individual pockets of hydrogen and helium gas come together under the influence of gravity until hydrogen undergoes nuclear fusion and begins to burn. Stars vary in size, temperature (or colour) and brightness. These features and the 'life cycles' of stars are determined by their mass. Three main mass categories are identified.

1. High mass stars (much more massive than the Sun, the largest, hottest and brightest stars, blue or white in colour).

2. Low mass stars (much less massive than the Sun, the smallest, coolest and dimmest stars, orange, red or brown in colour).

3. Intermediate mass stars (about the mass of the Sun, yellow in colour).

High mass stars burn their hydrogen rapidly, have the shortest lives, and die catastrophically in supernovae explosions. A supernova may leave behind a remnant neutron star (an object about 20km in diameter composed almost entirely of neutrons), a pulsar (a rapidly rotating neutron star) or a black hole (an object probably smaller than a neutron star with a gravity so strong that not even light can escape). Low mass stars burn their hydrogen slowly, have the longest lives and die by simply burning up their fuel and leaving behind white dwarfs (remnant stars about the size of the Earth). Intermediate mass stars, like the Sun itself, end their lives by passing through a red giant stage before leaving behind white dwarfs. Red giants form when a star's main source of fuel, hydrogen, is used up and other elements such as helium begin to burn. This causes the star's outermost layers to expand and cool, thus appearing red. Our own Sun, an intermediate mass star, will enter its red giant stage a few billion years from now. As it expands to many times its original size, it will engulf all of the inner planets including the Earth.

How stars end their lives is important in terms of how matter other than hydrogen and helium is produced and released into space to be incorporated into later generations of stars. In the process of forming stars from nebulae rich in hydrogen and helium gas and other matter such as ice and dust, the formation of planets, moons and other objects is also possible.

PRACTICAL TASK PRACTICAL TASK **PRACTICAL TASK** PRACTICAL TASK

Concept mapping is one of many techniques which allows the 'structure' of children's ideas to be explored. The maps themselves can reveal relationships and hierarchies between concepts (major and subordinate elements), propositions (meaningful connections between elements in words), cross-links (meaningful connections between elements as lines or arrows) and examples (specific objects or events). Use the concept map in Figure 13.1 to provide one example of each of the four features described. Are the elements hierarchical or not? Are the elements interconnected or not? Are there any factual inaccuracies? How would you assess this child's knowledge of the Universe, galaxies and stars?

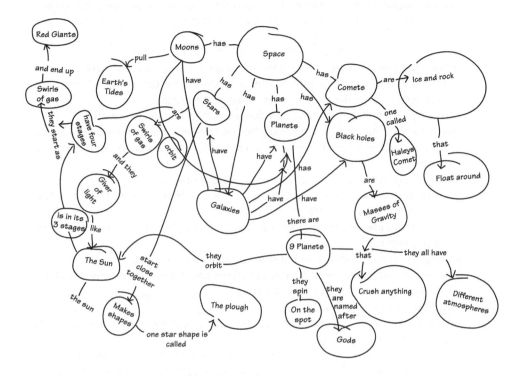

Figure 13.1. Child's concept map of the Universe, galaxies and stars

The Solar System

The Sun

The **Solar System** is the name given to our nearest star, the Sun, together with its family of nine planets, their moons, and other objects including asteroids, comets and meteoroids (see Figure 13.2). The Sun is by far the most important feature of the Solar System, accounting for about 99.8 per cent of its entire mass. With a diameter of about 1.4 million kilometres, the Sun is about 100 times the size of the Earth. The Sun and the rest of the Solar System formed about 4.6 billion years ago from a rotating nebula of gas, ice and dust, a nebula enriched by the deaths of nearby stars. The Sun's composition is very much dominated by hydrogen (74 per cent) and helium (25 per cent) with heavier elements making up the remainder (1 per cent). Like all other typical stars, the Sun burns hydrogen in its core. The temperature within the core is about 15 million °C. The temperature nearer the surface is a mere 6 000°C. The Sun is the only object within the Solar System which produces its own light.

The relative ease with which children learn about the Solar System has been noted by Sharp and Kuerbis (2005). They suggest that many children may actually use their knowledge and understanding of the Solar System to locate, contextualise and underpin work associated with the Earth–Sun–Moon System. If this is actually true, then there is a strong case for teaching about the Solar System either before or alongside such things as day and night, the seasons and the phases of the Moon. Traditional views of how children learn science are being challenged all the time. In some areas of science, and in the later primary years at least, there is clear evidence to suggest that children's abilities and levels of attainment have been underestimated.

As the Sun is a typical star, it does not have a solid surface; it has layers of gas of different density. The outermost three are:

- the photosphere;
- the chromosphere;
- the corona.

The photosphere, the Sun's visible 'disk', has a granulated appearance owing to the convection of hot gases from deep within the solar interior. Cooler areas of the photosphere which appear black are known as sunspots. Sunspots can reach up to tens of thousands of kilometres across.

Sunspot studies reveal that the Sun rotates once every 25 days at its equator but once every 35 days at its poles. The chromosphere and the corona can be seen only during solar eclipses. Jets of glowing gas called prominences, and flares of high energy radiation and atomic particles, the Solar Wind, extend from the Sun's surface layers and reach far out into space.

Planets and their moons

For more information see the section on gravity in Chapter 10.

The **planets** of the Solar System are spherical or nearly spherical in shape. They move around the Sun in nearly circular or elliptical orbits, in the same direction (anticlockwise when viewed from 'above') and in pretty much the same plane. The planets are held in orbit around the Sun by the Sun's gravitational pull. The planets are classified into two groups:

- the inner planets (Mercury, Venus, Earth and Mars);
- the outer planets (Jupiter, Saturn, Uranus, Neptune and Pluto).

The inner planets are also referred to as the terrestrial or rocky planets. The outer planets, with the exception of Pluto whose nature and origin remain uncertain, are also referred to as the Jovian planets or gas giants. While Mercury, Venus, Earth, Mars, Jupiter and Saturn were all known to ancient astronomers, Uranus, Neptune and Pluto were discovered only in more recent times. There has been debate as to whether Pluto is actually a planet (see the Reflective Task later in this chapter).

RESEARCH SUMMARY RESEARCH SUMMARY RESEARCH SUMMARY **RESEARCH SUMMARY**

Studies from around the world, including those by Ehrlén (2008), indicate that many children start school with ideas about the shape of the Earth that are anything but scientific. Some think the Earth is flat, some think it hollow and covered with a dome-like sky, and some think that two Earths exist, the one they live on and another which they often hear talked about! Ehrlén interviewed chldren aged 6–8 about their concept of the Earth as a sphere. Many children had problems with the astronomical model of the Earth but they seemed able to hold their common-sense framework about the planet and yet at the same time use a globe to talk about orientation on a sphere.

Details of the planets and their **moons** are provided below. Note that distances within the Solar System are not measured in light years but in astronomical units (AU). 1AU is about 93 million miles or 150 million kilometres, the average distance between the Sun and the Earth.

Mercury is the nearest planet to the Sun. At a distance of about 58 million kilometres or 0.4AU, and travelling at about 48 kilometres per second, Mercury orbits the Sun once every 88 days. Mercury also rotates on its axis once every 59 days. With a diameter of 4878 kilometres, Mercury is about 0.4 times the size of the Earth. Most of Mercury's surface is heavily cratered and of volcanic origin. The largest feature observed on its surface is known as the Caloris Basin, a crater about 1300 kilometres across. With daytime highs of 420°C and night-time lows of –170°C, Mercury experiences the greatest temperature range of all the planets. Mercury has no permanent atmosphere and no moons.

Venus is the second planet from the Sun. At a distance of about 108 million kilometres or 0.7AU, and travelling at about 35 kilometres per second, Venus orbits the Sun once every 224.7 days. Venus also rotates on its axis once every 243 days but in the opposite direction to the majority of other planets. With a diameter of 12 102 kilometres, Venus is about 0.95 times the size of the Earth. The surface of Venus is cratered, mountainous and of volcanic origin. With highs of 430°C, Venus is the hottest planet in the Solar System. Venus has a dense atmosphere dominated by carbon dioxide with minor amounts of sulphuric acid. Venus has no moons.

Earth is the third planet from the Sun and unique in that it is currently the only planet known to support life. At a distance of about 150 million kilometres or 1.0 AU, and travelling at about 30 kilometres per second, the Earth orbits the Sun once every 365.25 days. The Earth also rotates on its axis once every 23 hours 56 minutes. The Earth's diameter is 12 756 kilometres. About 70 per cent of the Earth's surface is covered by water. Surface features on its land masses are wide and varied. Surface temperatures range from 60°C to –90°C. The Earth's atmosphere is dominated by nitrogen and oxygen with traces of other gases. The Earth has one moon, *the Moon*, its only natural satellite.

Mars is the fourth planet from the Sun. At a distance of about 228 million kilometres or 1.5AU, and travelling at about 24 kilometres per second, Mars orbits the Sun once every 687 days. Mars also rotates on its axis once every 24 hours 37 minutes. With a diameter of 6786 kilometres, Mars is about 0.5 times the size of the Earth. Surface features on Mars are wide and varied and include mountains, deserts, canyons, volcanoes and polar caps of frozen carbon dioxide. Running water may have been present in the past. Two particularly impressive features are Olympus

Figure 13.2 The Solar System

Mons, the largest volcano in the Solar System (now extinct), and Valles Marineris, a canyon over 4000 kilometres long and up to 7 kilometres deep. Surface temperatures range from 20°C to –140°C. Mars has a thin atmosphere dominated by carbon dioxide. Mars has two moons, Phobos (22 kilometres across) and Deimos (13 kilometres across).

Jupiter is the fifth planet from the Sun. At a distance of about 778 million kilometres or 5.2AU, and travelling at about 13 kilometres per second, Jupiter orbits the Sun once every 11.86 years. Jupiter also rotates on its axis once every 9 hours 48 minutes. With a diameter of 142 980 kilometres, Jupiter is about 11.2 times the size of the Earth and the largest planet in the Solar System. As a gas giant, Jupiter is dominated by cloud layers of hydrogen, helium and other gases. In its upper levels, these appear colourful and strongly banded. Cloud top temperatures reach as low as –110°C. Jupiter's most prominent feature is its Great Red Spot, an anticylonic weather system reaching up to 40 000 kilometres across. Jupiter has narrow and faint rings and over 60 moons, the largest of which are Io, Europa, Ganymede and Callisto.

Saturn is the sixth planet from the Sun. At a distance of about 1427 million kilometres or 9.5AU, and travelling at about 10 kilometres per second, Saturn orbits the Sun once every 29.46 years. Saturn also rotates on its axis once every 10 hours 15 minutes. With a diameter of 120 540 kilometres, Saturn is about 9.5 times the size of the Earth. Like Jupiter and the other gas giants, Saturn is dominated by cloud layers of hydrogen, helium and other gases. In its upper levels, these appear pale and weakly banded. Cloud top temperatures reach as low as –180°C. Saturn's most prominent and spectacular feature is its extensive ring system, which extends into space up to 1 million kilometres around the planet. Saturn has over 40 moons, the largest of which is Titan.

Uranus, the seventh planet from the Sun, was discovered in 1781. At a distance of about 2871 million kilometres or 19.2AU, and travelling at about 7 kilometres per second, Uranus orbits the Sun once every 84.01 years. Uranus also rotates on a very tilted axis once every 17 hours 4 minutes. With a diameter of 51 118 kilometres, Uranus is about 4.0 times the size of the Earth. Like Jupiter and the other gas giants, Uranus is dominated by cloud layers of hydrogen, helium and other gases. In its upper levels, these appear featureless but blue owing to the presence of methane. Cloud top temperatures reach as low as –216°C. Uranus has faint rings and over 25 moons, all of which are named after Shakespearean characters.

Neptune, normally the eighth planet from the Sun, was discovered in 1846. At a distance of about 4497 million kilometres or 30.0AU, and travelling at a little over 5 kilometres per second, Neptune orbits the Sun once every 164.80 years. Neptune also rotates on its axis once every 16 hours 6 minutes. With a diameter of 49 528 kilometres, Neptune is about 3.99 times the size of the Earth. Like Jupiter and the other gas giants, Neptune is dominated by cloud layers of hydrogen, helium and other gases. In its upper levels these appear faintly banded and blue owing to the presence of methane. Neptune's one-time most prominent feature, the Great Dark Spot, an anticyclonic weather system reaching up to 20 000 kilometres across, no longer exists. Neptune is the windiest place in the Solar System with recorded gusts of up to 2000 kilometres per hour. Cloud top temperatures reach as low as –216°C. Neptune has faint rings and over 10 moons, the largest of which is Triton.

Pluto, normally the ninth and furthest planet from the Sun, was discovered in 1930 (see Reflective Task below). At a distance of about 5914 million kilometres or 39.50AU, and travelling at almost 5 kilometres per second, Pluto orbits the Sun once every 248.5 years. Pluto's orbit is so much more elliptical than the other planets that for about twenty years of its orbital period it is closer to the Sun than Neptune (e.g. between 1979 and 1999). Pluto's orbit is also more inclined relative to the orbital planes of the other planets. Pluto rotates on its axis once every 6.4 days. With a diameter of 2300 kilometres, Pluto is about 0.2 times the size of the Earth and the smallest planet in the Solar System. Surface temperatures on Pluto may reach as low as –220°C. Pluto has three moons, the largest of which is Charon.

REFLECTIVE TASK

Is Pluto a planet? This question has taxed astronomers for some considerable time. Pluto is very small (smaller than some moons), its orbit is highly elliptical and tilted with respect to the other planets (so much so it was recently the 'eighth' rather than the 'ninth' planet), and its composition is thought to be more 'terrestrial' and therefore out of place with the other inner planets (it usually lies beyond the four major gas giants). In 2006, the International Astronomical Union voted to downgrade Pluto to a 'dwarf planet' alongside the largest object in the asteroid belt, Ceres, and a few other recently discovered bodies of similar size. While this example serves to illustrate the sometimes transient nature of scientific knowledge (though only by way of a definition), what do we teach children in schools? How long will this knowledge, if it remains uncontested, take to filter through the education system? Will the astronomy component of the national science curriculum be changed and when? Will there be confusion with national tests? What do you think?

Other objects

Other objects within the Solar System, 'debris' not involved in the formation of planets and their moons, include:

- asteroids;
- comets;
- meteoroids.

Most **asteroids** occupy a region of space between Mars and Jupiter known as the asteroid belt. Belt asteroids are relatively small, irregular or spherically shaped lumps of rock that, like planets, orbit the Sun. The largest, Ceres, is about 900 kilometres across (see Reflective Task). **Comets** are essentially chunks of ice and other material that occupy a region of space well beyond the outer planets known as the Oort Cloud. Some comets travel in highly elliptical paths that bring them towards the inner planets as they orbit the Sun. As a comet approaches the Sun, its icy matter begins to vaporise, producing a large head or coma and a tail millions of kilometres in length. The tail is pushed away from the Sun by the Solar Wind. Most **meteoroids** are particles of dust and rock fragments derived from comets. When a meteoroid enters the Earth's atmosphere it begins to burn up, producing a trail known as a meteor or a shooting star. Meteoroids that survive and reach the Earth's surface are known as meteorites.

PRACTICAL TASK PRACTICAL TASK **PRACTICAL TASK** PRACTICAL TASK

Models of the Solar System are frequently displayed along classroom walls and school corridors or, alternatively, hung from ceilings as mobiles. Displayed like this, the scale of the Solar System is often overlooked and misrepresented. Using a scale of 1 metre to 1 million kilometres, calculate how big and how far apart the Sun and the planets would be. Why is using one scale for both size and distance problematic? Use the information presented in this chapter to compile a database for the planets. Interrogate the database for any patterns and relationships.

The Earth–Sun–Moon System

Day and night

The parts of the Earth's surface that face towards the Sun are lit by it and experience day, while the parts of the Earth's surface that face away from the Sun are in darkness and experience night. The day and night cycle is caused by the rotation of the Earth about its axis. Because the Earth always rotates the same way (anti-clockwise looking 'down' on the North Pole), the Sun always appears to 'rise' at dawn generally towards the east and 'set' at dusk, generally towards the west. The Earth rotates once every 23 hours 56 minutes. As it rotates, however, it is also moving in orbit around the Sun. As a result, the time from one 'sunrise' to the next, the day and night cycle, is 24 hours.

Seasonal variations in where the Sun actually 'rises' and 'sets' and changes in the length of daylight hours throughout the year are due to the tilt of the Earth's axis. The Earth's axis tilts at 23.5° from the vertical relative to the plane of the Earth's orbit around the Sun. In June, the northern hemisphere is tilted towards the Sun while the southern hemisphere is tilted away (see Figure 13.3). The effects are quite dramatic. In the UK:

- the Sun 'rises' in the northeast, 'passes' at its highest in the sky to the south, and 'sets' in the northwest;
- days are long and nights are short (the Sun spends up to 18 hours above the horizon).

RESEARCH SUMMARY RESEARCH SUMMARY **RESEARCH SUMMARY RESEARCH SUMMARY**

Collins and Simpson (2007) explore children's and their teachers' concepts of the Earth and Moon system. They use uncertainty about the orbit of the Moon to speculate about the way we learn complex and apparently contradictory concepts. They also examine how the language we use reinforces misconceptions.

In December, the opposite occurs. The northern hemisphere is tilted away from the Sun, while the southern hemisphere is tilted towards it. In the UK:

- the Sun 'rises' in the southeast, 'passes' at its lowest in the sky to the south, and 'sets' in the southwest;
- days are short and nights are long (the Sun spends as few as 6 hours above the horizon).

During March and September, the northern and southern hemispheres are tilted neither towards nor away from the Sun.

- The Sun 'rises' due east and 'sets' due west.
- Day and night are of equal duration (the Sun spends 12 hours above the horizon).

Standing on the surface of the Earth, its shape and the fact that it rotates about a tilted axis are not at all obvious. Clues that the Earth is at least curved and not flat come from observing ships at sea 'disappearing' over the horizon and the shadow of the Earth on the Moon during a lunar **eclipse**. Images of the Earth from space provide the most convincing and most easily accessible evidence. Changes in the length of daylight hours and changes in the height of the Sun in the sky throughout the course of a year provide at least some evidence that the Earth's axis is tilted and not upright.

The seasons

In addition to causing variations in where the Sun 'rises' and 'sets' and the length of daylight hours throughout the year, the tilt of the Earth's axis relative to the plane of its orbit around the Sun causes the **seasons** (see Figure 13.4). In the UK, the year-long cycle of seasons includes:

- spring (taken by astronomers to begin at the time of the spring equinox on 21 March, when day and night are of equal duration);
- summer (taken by astronomers to begin at the time of the summer solstice on 21 June, the longest day);
- autumn (taken by astronomers to begin at the time of the autumnal equinox on 22 September, when day and night are of equal duration);
- winter (taken by astronomers to begin at the time of the winter solstice on 21 December, the shortest day).

In June, the northern hemisphere is tilted towards the Sun and experiences summer, while the southern hemisphere is tilted away and experiences winter. As with day and night, the effects are quite dramatic. In the UK:

- days are long;
- the Sun 'rises' high above the horizon, so the Sun's rays reach the surface of the Earth at a high angle;
- the Earth is heated by the Sun for more than 12 hours and the Sun's heating effect is more efficient (summers are warm).

In December, the opposite occurs. The northern hemisphere is tilted away from the Sun and experiences winter, while the southern hemisphere is tilted towards the Sun and experiences summer. In the UK:

- days are short;
- the Sun remains low above the horizon so the Sun's rays reach the surface of the Earth at a low angle;
- the Earth is heated by the Sun for fewer than 12 hours and the Sun's heating effect is less efficient (winters are cold).

While the distance between the Sun and the Earth is about 150 million kilometres, there are times during its nearly circular or elliptical orbit when it is closer and times when it is further away. Interestingly, at **aphelion**, when the Earth is furthest away (152 million kilometres), the northern hemisphere is tilted towards the Sun and experiences summer. The opposite is true at **perihelion** when the Earth is closest to the Sun (147 million kilometres). While distance does make a difference to the amount of energy actually received from the Sun, the tilt of the Earth's axis is more significant.

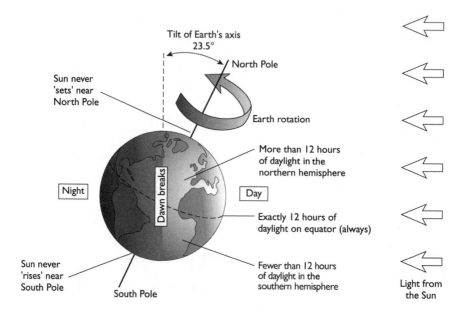

Figure 13.3 Day and night (June)

The phases of the Moon and eclipses

The Moon is the Earth's only natural satellite. At a distance of about 384 400 kilometres, and travelling at about 1 kilometre per second, the Moon orbits the Earth once every 27 days 7 hours. The Moon also rotates on its axis once every 27 days 7 hours, explaining why only the Moon's 'near' side is ever seen. With a diameter of 3476 kilometres, the Moon is about 0.25 times the size of the Earth. Its heavily cratered surface and 'seas' or plains of old lava flows are its most prominent features. The Moon has no permanent atmosphere. Surface temperatures range from 130°C in direct sunlight to –180°C in darkness. Between 1969 and 1972, the Moon was visited and landed on by American astronauts during six successful Apollo missions.

Over a period of one lunar month (29 days 12 hours), the Moon's shape appears to change in a regular and predictable way. A lunar month is longer than the Moon's rotation period by about 2 days because the Earth is moving in orbit around the Sun at the same time as the Moon is moving in orbit around the Earth. Like the Earth, the parts of the Moon's surface that face towards the Sun are lit by it and experience day (a lunar day), while the parts of the Moon's surface that face away from the Sun are in darkness and experience night (a lunar night). The apparent shapes or phases of the Moon (see Figure 13.5) are caused by the location of the Moon in its orbit around the Earth and the amount of the Moon's lit surface that can be seen by simply looking at it.

- From a new Moon, when the Moon is situated in orbit between the Earth and the Sun and not visible in the daytime sky, the Moon waxes or appears to 'grow' until it becomes full (the amount of visible surface increases through crescent, first quarter and gibbous phases).
- From a full Moon, when the Moon is situated in opposition to the Earth and the Sun and at its most visible in the night-time sky, the Moon wanes or appears to 'shrink' until it is not visible again (the amount of visible surface decreases through gibbous, last quarter and crescent phases).

Total solar eclipses occur when the Moon is exactly positioned between the Sun and the Earth, and the Earth passes through the Moon's shadow. Solar eclipses are spectacular events. While the Sun is about 400 times bigger than the Moon, it is also about 400 times further away from the Earth. Both the Sun and the Moon therefore appear the same size in the sky. The Moon covers the Sun's 'disk' perfectly and blocks out its light. Solar eclipses can be seen only during the day. A total lunar eclipse occurs when the Earth is exactly positioned between the Sun and the Moon, and the Moon passes through the Earth's shadow. Lunar eclipses can be seen only at night.

PRACTICAL TASK PRACTICAL TASK PRACTICAL TASK PRACTICAL TASK

Tracking the apparent movement of the Sun across the sky can be undertaken indirectly by drawing around children's shadows with chalk in the playground and observing how they change over time. Consider yourself doing this with your class on a cloudless day in June and again on a cloudless day in December. Apply your knowledge and understanding of the Earth–Sun–Moon System to predict how the shadows would appear when measured first thing in the morning, at lunchtime and last thing in the afternoon on each day. Sketch your ideas, showing the shadows in relation to the position of the Sun side by side. How are the shadows similar and how are they different? How could you extend this work using a model of the Earth–Sun–Moon System involving a globe and an overhead projector?

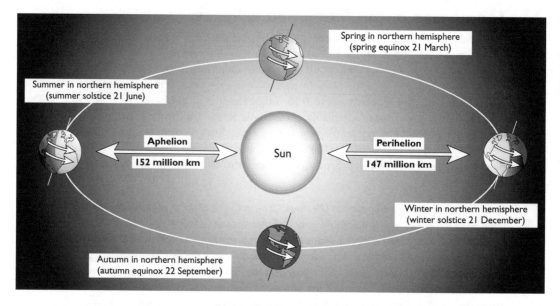

Figure 13.4 The seasons

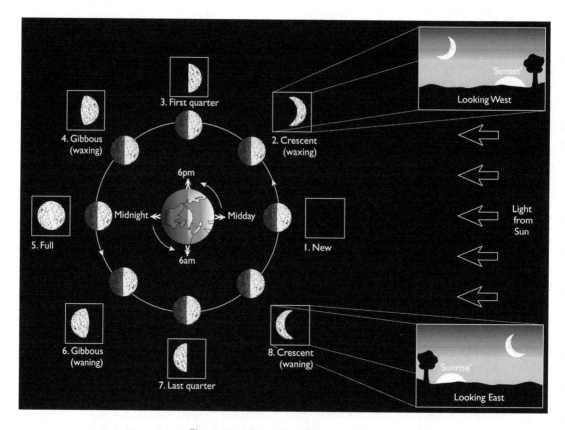

Figure 13.5 The phases of the Moon

A SUMMARY OF **KEY POINTS**

> The Universe, all the matter, radiation and space that there is, has been expanding since the time of the Big Bang about 14 billion years ago.

> Galaxies are assemblages of stars, nebulae and other interstellar material and occur in clusters and superclusters.

> Stars are balls of hot, glowing gas born out of the nebulae within galaxies.

> The Solar System is the name given to the Sun together with its family of planets, their moons, and other objects, including asteroids, comets and meteoroids.

> The Sun is a star; the Earth is a planet; the Moon is the Earth's only natural satellite.

> The day and night cycle is caused by the rotation of the Earth once every 24 hours.

> Variations in daylight hours, the apparent position of the Sun in the sky, and the seasons are caused by the tilt of the Earth's axis.

> The Earth orbits the Sun once a year.

> The phases of the Moon are caused by the amount of the Moon's lit side that can be seen as it orbits the Earth about once a month.

> Eclipses occur when the Earth, the Sun and the Moon are in perfect alignment.

M-LEVEL EXTENSION > > > > M-LEVEL EXTENSION > > > >

Although this area of science contains some very difficult concepts for children to understand, many enjoy the challenge of finding out about the elements of the Solar System, especially the planets. If you plan to link this work with developing children's skills in using information books or research skills using the internet, think about how you could differentiate the activities. If setting the task of researching the planets, for example, consider asking a more able group to present the arguments for and against Pluto being considered as a planet, and the history of its discovery and reclassification. While some groups could find basic facts about planets, moons, comets, asteroids and meteoroids, others could research the phases of the Moon or the story of manned and unmanned space exploration and how we know about the Solar System and beyond. Look back at the research summaries and reflect on what should be introduced to children and when.

REFERENCES REFERENCES **REFERENCES** REFERENCES REFERENCES

Collins, R. and Simpson, F. (2007) Does the moon spin? *Primary Science Review*, 97, 9–11.

Ehrlén, K. (2008) Children's understanding of globes as a model of the earth: a problem of contextualising. *International Journal of Science Education*, 30(2), 221–38.

Kibble, B. (2002) Misconception about space? It's on the cards. *Primary Science Review*, 72, 5–8.

Sharp, J. G. and Grace, M. (2004) Anecdote, opinion and whim: lessons in curriculum development from primary science education in England and Wales. *Research Papers in Education*, 19(3), 293–321.

Sharp, J. G. and Kuerbis, P. (2005) Children's ideas about the Solar System and the chaos in learning science. *Science Education*, 90(1), 124–47.

FURTHER READING FURTHER READING **FURTHER READING** FURTHER READING

DfE (2011) *Teachers' Standards*. Available at www.education.gov.uk/publications.

Dorling Kindersley. *Eyewitness Encyclopedia of Space and the Universe*. London: Dorling Kindersley Multimedia. An excellent, information-packed, interactive CD-ROM with video clips and spoken text.

Hollins, M. and Whitby, V. (2001) *Progression in Primary Science: a Guide to the Nature and Practice of Science in Key Stages 1 and 2*. London: David Fulton. Provides useful information on teaching strategies and, as the title suggests, children's progression in all areas of National Curriculum science.

Sharp, J. (ed.) (2004) *Developing Primary Science*. Exeter: Learning Matters. Provides useful information on all aspects of science education.

Self-assessment questions

Chapter 2 – Functioning of organisms: green plants

1. What common names are given to the two major groups or classes of flowering plants?

2. What is the significance of the mnemonic 'Mrs Gren'?

3. How are plant and animal cells different? (You may wish to refer to Figures 2.2 and 3.2.)

4. Using your knowledge and understanding of light and colour, explain why green plants are green.

5. What are the main functions of the following?
   • Roots
   • Stems
   • Leaves
   • Flowers

6. What is photosynthesis, why is it important and where does it normally take place?

7. Which factors limit the rate at which photosynthesis takes place?

8. How do phloem and xylem cells differ from other plant cells?

9. Why can seeds germinate and grow in the dark but fully-grown plants cannot?

10. Why do you think that the widespread use of herbicides, insecticides and fungicides causes so much concern?

11. Sketch a flowering plant. Label the roots, stem, leaves and flowers and describe the main function of each of these. Produce a sketch of a flower in more detail showing the male and female reproductive organs. In a flowering plant, how does sexual reproduction take place?

12. What are the limiting factors for seed germination and photosynthesis? Design as many different investigations or experiments to test each one as you can. Begin by predicting outcomes and stating the hypotheses that you wish to test.

Chapter 3 – Functioning of organisms: humans and other animals

1. How are humans and other animals different from other living organisms?

2. What words do the initial letters of each life process spell out and why are these significant?

3. What is DNA?

4. What type of joint are knee and hip joints?

5. Why do skeletal muscles work in antagonistic pairs?

6. Give three examples of reflex action.

7. What is unusual about the pulmonary artery and the pulmonary vein?

8. Why is it important to clean teeth at least twice per day, using an approved toothpaste and a good brush?

9. What is special about the male and female gametes or sex cells?

10. Why are antibiotics limited in the treatment of colds and flu?

11. What must you do before bringing any living organisms (other than humans) into a primary classroom?

12. Produce sketches of the human skeletal, muscular, circulatory and digestive systems, labelling as many of the different 'parts' as you can and annotating them with notes outlining how each system works. Compare your sketches with the illustrations provided in the chapter. How accurate were you about size, shape, location and function?

13. List the main vertebrate classes within Phylum Chordata. What are their defining characteristics? List the main invertebrate classes within Phylum Arthropoda. What are their defining characteristics? Produce a formal biological classification scheme that shows the relationship between these groups. Where would you place annelids and molluscs?

Chapter 4 – Continuity and change

1. Judging by the genes and species names, how closely related are the blue tit (Parus caeruleus) and the coal tit (Parus ater) compared with the carrion crow (Corvus corone corone) and the hooded crow (Corvus corone cornix)?

2. In Australia and New Zealand, there are many endemic species, i.e. species that are found nowhere else on Earth. Why is this?

3. Are there more chromosomes in brain cells or skin cells?

4. Are there more chromosomes in a human egg just emerging from the ovary or in a fertilised egg? Explain your answer.

5. Two normal parents, both of whom have the recessive gene for sickle cell disease, have two children, both of whom have the disease. The parents think that they would be unlikely to have a third child affected by the disease. Is this the case?

6. In two sentences, summarise why you are for or against the genetic modification of crops.

7. Of the five groups of vertebrates, which developed first and which are the most recent innovations?

8. In order of specificity, what are the features that distinguish humans from the other great apes?

9. Summarise the evolutionary pressures that led to the development of whales

and dolphins. Explain why they continue to breathe through lungs rather than gills.

10. Comment on the idea that humans are evolving towards bodily perfection.

Chapter 5 – Ecosystems

1. Which terms could be used to describe the points in the following food chains?
   - Sun → grass → sheep → human
   - Sun → courgette → human

2. How many individual food chains make up the food web illustrated in Figure 5.2?

3. What impact would a rapid decline in the population of blue tits have on the food pyramid shown in Figure 5.3?

4. Using the knowledge that you gained from reading the chapter, and the freshwater food chains below, create a food web for freshwater animals.
   - phytoplankton → daphnia → phantom midge → roach → pike
   - phytoplankton → cyclops → stickleback → pike
   - phytoplankton → daphnia → stickleback → perch → pike
   - phytoplankton → cyclops → perch → pike
   - phytoplankton → daphnia → pike

5. Using the knowledge that you gained from reading the chapter, and the food chains below, create a food web for garden animals.
   - dead leaves, fungi and bacteria → mite → ant → centipede → hedgehog
   - dead leaves, fungi and bacteria → springtail → centipede → hedgehog
   - dead leaves, fungi and bacteria → vegetation → earthworm → hedgehog
   - dead leaves, fungi and bacteria → vegetation → slug → centipede → hedgehog
   - vegetation → slug → hedgehog
   - dead leaves, fungi and bacteria → earthworm
   - dead leaves, fungi and bacteria → mite → snail → violet ground beetle → hedgehog
   - vegetation → snail → violet ground beetle → hedgehog

6. Use labels to identify each point in the food chains in Question 5, for example primary producer, primary consumer, secondary consumer, tertiary consumer, or predator, prey, or herbivore, carnivore, omnivore.

Chapter 6 – Materials

1. How are protons and neutrons the same/different?

2. What is the atomic mass of sulphur, and how many protons and neutrons are in its nucleus?

3. What is the difference between ionic and covalent bonding?

4. How is solid gold chemically different from liquid gold?

5. In terms of electrons and the periodic table:
   a) What do argon, helium and neon have in common?

b) What do sodium and potassium have in common?

c) How would an oxygen ion be represented?

d) Show the covalent bonding in N_2 using a dot and cross diagram.

e) What kind of bonding will there be in magnesium oxide (MgO)? Show in diagrammatic form what a single molecule would look like.

f) What kind of bonding will there be in ethanol (C_2H_5OH)? Show in diagrammatic form what a single molecule would look like.

Chapter 7 – Particle theory and the conservation of mass

1. What happens when you add tea leaves to water?

2. Where does the oxygen come from in this chemical reaction?

iron + oxygen = iron oxide

3. How is the process of burning similar to what happens when living things respire?

4. Where does the oxygen come from when a material is burned?

5. What happens when table salt dissolves in water?

6. What would happen to the cap of a full bottle of water that is kept in the freezer? Explain your answer.

Chapter 8 – Electricity and magnetism

1. Give two reasons why fuse wire has a higher resistance than flex.

2. What is likely to happen if an extension cable made from thin flex is used to supply a heater?

3. Why is resistance a problem for electrical supply companies?

4. Would you expect two wires side by side to conduct electricity better than a single wire? Explain your answer.

5. Why does a variable resistor get warm?

6. Predict the brightness of the bulbs in the circuits below, if a 3 volt battery is used and the bulbs are as follows:
Bulb (a) 6 volts/0.2 amps ... this bulb has a resistance of 30 Ω
Bulb (b) 2.5 volts/0.2 amps ... this bulb has a resistance of 12.5 Ω
Bulb (c) 2.5 volts/0.1 amps ... this bulb has a resistance of 25 Ω

i) A series circuit with two Bulbs (b)
ii) A series circuit with Bulb (b) and Bulb (c)
iii) A parallel circuit with Bulb (a) in one pathway and Bulb (b) in the other

7. Look back at the Practical task in the section on series and parallel circuits. If the second pathway does not have any chair in it, why does this act like a short circuit?

8. Calculate the time that it would take:
a) A 100W bulb to use 1 kW hour of electricity
b) A 20W fluorescent tube to use 1 kW hour of electricity

9. a) How many joules of energy will an 8kW shower use in 10 minutes?
    b) Will it use less energy than heating a 3kW kettle for an hour?

10. Magnets attract tin cans. Does that mean that tin is magnetic?

11. Why do bulbs in series circuits glow less brightly than the same bulbs in a parallel circuit?

12. Which has the higher resistance: a 100W bulb or a kettle? Explain your answer.

13. Explain why the wire flex leading to a kettle is thicker than the flex leading to a table lamp.

Chapter 9 – Energy

1. How much work is done if a pram is pushed with a force of 60N for 50 metres?

2. How much work is done keeping a 500 W security lamp on for 2 hours?

3. Will a red-hot drawing pin have more heat energy than a large pan of boiling water? Explain your answer.

4. What form of energy is associated with lying on a mattress?

5. What is the liquid that drips out of car exhausts on cold days?

6. If digestion takes place in the gut, where does respiration take place?

7. Why does the growth of birds, animals and arthropods account for such a small portion of their food intake?

8. Thinking about converting energy in our bodies, how much chocolate would you need to replace the energy used in 8 hours of sleeping?

9. How does a thermos flask keep tea hot for long periods?

10. Would a fridge with the door left open cool a perfectly insulated room?

11. Imagine that we were able to collect all of the gases given off by a burnt candle. Would the mass of the gases be less than the original candle, more than the original candle or the same as the original candle? Explain your answer.

12. When plants photosynthesise, what chemical bonds are they breaking in the materials that they use?

Chapter 10 – Forces and motion

1. Bubbles float upwards in water. Can you think of any examples of things that float upwards in air?

2. Imagine that the Moon is being swung around the Earth very much like a ball on the end of a rope. What provides the force to enable the Moon to constantly change its direction (i.e. to accelerate)?

3. Which of the possible methods of reducing friction are used by an ice skater?

4. Draw the following simple situations involving a ball or a brick and use arrows to show:

- the relative size of all of the forces acting on each ball or brick;
- the direction of each force.

a) A ball resting on a table
b) A brick resting on a slope
c) A ball just after it has been dropped from a hand
d) A ball just after it has been thrown upwards by a hand
e) A ball floating on water
f) A brick as it drops through water
g) A brick resting on the bottom of a container of water

Chapter 11 – Light

1. Think of some examples of where we see things almost instantaneously but hear the sound that they make a little later.

2. What kind of energy is changed to light when a battery lights a bulb?

3. What happens when you cover one eye for a while and then look at both pupils in a mirror?

4. How could you make a totally black shadow?

5. Which colour is mostly absorbed by orange paint?

6. What light would you get on an overhead projector screen if you overlapped red and blue filters on it?

7. What are the three primary colours of light?

8. Describe the ways in which you might show children that light travels in straight lines.

9. Draw a picture to show the kind of shadow formed by a person standing between a neon striplight and a wall. How would this shadow be different if the light was a small point source like the Sun?

10. Explain why a blue ball looks blue.

Chapter 12 – Sound

1. What happens to the vocal chords in the voice box when you whisper?

2. On an oscilloscope:
   a) What would the display show for a loud high-pitched whistle?
   b) What kind of sound would make a display that showed low and well spaced out waves?

3. If there was a large explosion in outer space near to our planet, what would we see and what would we hear?

4. What is the amplitude of a vibration and how does it affect the loudness of a sound?

5. Someone wants to play the first three notes of *Three Blind Mice*. (These will be in descending pitch.) Draw the bottles in the order in which they should be struck.

6. Describe two ways in which you could help children to understand that sound travels better in solids than in air.

7. Figure 12.6 is a child's drawing of how a sound is heard through a door. List the child's possible misconceptions (alternative frameworks) regarding sound. List the ideas that the child does correctly understand about sound.

Chapter 13 – The Earth and beyond

1. Which force will determine the ultimate fate of the Universe?

2. Which observational feature of galaxies do you think first suggested that they might be rotating?

3. Stars are born, they live and then they die. Using the knowledge and understanding of life processes gained from earlier chapters, what do you think are the advantages and disadvantages of this 'life cycle' analogy?

4. If the Sun is the only object within the Solar System to produce its own light, how are we able to see the planets, their moons and the other objects around?

5. Why is it dangerous to look at the Sun without proper eye protection?

6. Why do you think that ancient astronomers believed in a Universe with the Earth at its centre?

7. List the six major components of the Solar System.

8. On any given day of the year, who would see the Sun 'rise' first: people living in New York or London?

9. How long is a year and how do you think that it is measured?

10. How long is a 12-month lunar year in days and why do you think that lunar years are not the best timekeepers?

11. New Moons and full Moons occur within 29-day,12-hour lunar cycles but eclipses only happen once or twice per year. Why do eclipses not occur more often?

12. Produce your own concept map of the Universe, galaxies and stars. Are the elements hierarchical or not? Are the elements interconnected or not? Using the text of the chapter, check for any factual inaccuracies. How would you assess your own knowledge of the Universe, galaxies and stars?

13. Draw a picture of the Solar System. Add as much detail as you can. How much can you remember (e.g. names, orbital periods, rotation periods, moons, rings, relative and absolute sizes, relative and absolute distances, and features)?

14. Sketch how the Earth–Sun–Moon System would look in June. Which hemisphere is experiencing summer and why? Describe the Sun's location and apparent movement in the sky from 'sunrise' to 'sunset' as viewed from the UK. Assume that the Moon is in its full Moon phase. Where would it appear in your sketch? When would it be visible from the UK?

Answers to self-assessment questions

Chapter 2 – Functioning of organisms: green plants

1. The monocotyledonous plants (Liliidae) and dicotyledonous plants (Magnoliidae) are better known simply as monocots and dicots.

2. Mrs Gren is a mnemonic formed from the first letters of the life processes as they are presented here. Mnemonics are useful memory aids.

3. Plant and animal cells have many similarities but differ in certain key respects. Animal cells do not have cellulose cell walls, chloroplasts or large, permanent vacuoles.

4. Chlorophyll reflects the green component of visible or white light. Most of the energy absorbed is from the blue and red ends of the spectrum.

5. Roots anchor plants. Stems hold plants upright, keep leaves and flowers spread out, store starch and provide links between roots and shoots. The leaves of plants are where photosynthesis takes place. Flowers are reproductive structures.

6. Photosynthesis is the process by which plants make their own food. Without the energy from food, plants, like all other living organisms, would soon die. Photosynthesis takes place mostly in the leaves of plants though all green tissues are capable of photosynthesis.

7. The rate of photosynthesis is affected by light intensity, temperature, the concentration of carbon dioxide in air and water availability.

8. Phloem cells are connected together by perforated end walls or sieve plates. They have no organelles including a nucleus and no vacuoles. Xylem cells are dead. Their cell walls often contain lignin, which provides support.

9. Seeds germinate in the dark because they are able to draw on their own stored food reserves. In dicot plants, for example, food is stored in each of the two cotyledons. Fully grown plants need light in order to photosynthesise.

10. Herbicides, insectides and fungicides contain chemicals which not only kill off the organisms causing problems but many others too. They are not always selective. Dead or alive, organisms sprayed with pesticides can find their way into and contaminate different trophic levels of food chains. Some organisms eventually become resistant to chemicals.

11. Sketches and labelling will vary. Check against text and chapter illustrations. Reproduction takes place in five well defined stages: pollination, fertilisation, seed formation, seed dispersal and germination.

12. Seed germination can be affected by water availability, temperature and oxygen but not light. Photosynthesis can be affected by light, temperature, carbon dioxide concentration and water availability. Investigations or experiments may vary. If seeds do not need light to germinate, for example, this can be tested easily by germinating them in suitable locations ('Do seeds need light to germinate?'). Investigations or experiments with oxygen and carbon dioxide are generally not suitable for the primary classroom.

Chapter 3 – Functioning of organisms: humans and other animals

1. All means of classifying organisms are fraught with difficulties and there are no easy solutions. In addition to the features mentioned, human and other animal cells have no cellulose cell walls, chloroplasts or large, permanent vacuoles like the cells of plants. Protoctists are usually simple, unicellular organisms. Monerans are also usually simple, unicellular organisms but prokaryotic which means that their cells have no well-defined nucleus. Fungi are made up of microscope threads or hyphae rather than true cells. Individual hyphae may contain more than one nucleus.

2. They spell out Mrs Gren. Mrs Gren is a useful mnemonic or memory aid.

3. DNA is short for deoxyribonucleic acid. DNA is a large molecule arranged in the shape of a double helix. In the nucleus of a human cell, DNA occurs coiled up in structures called chromosomes. The nuclei of most human cells usually contain 23 pairs of chromosomes.

4. Both are freely movable joints. Knee joints are hinge joints. Hip joints are ball and socket joints.

5. Skeletal muscles can contract and pull on bones only when stimulated by nerve impulses. There are no corresponding nerve stimuli to relax them.

6. Blinking, coughing, sneezing and limb withdrawal are examples of reflex action.

7. Arteries usually carry oxygenated blood. Veins usually carry deoxygenated blood. The pulmonary artery and pulmonary vein do not follow this pattern.

8. Cleaning teeth in this way helps prevent the build-up of bacteria which feed on food particles and which produce acids which attack the hard enamel and dentine, causing tooth decay. Eventually, bacteria reach the pulp cavity and cause toothache. Cleaning teeth also helps prevent gum disease. The thin film of mucus, saliva, bacteria, food particles and other debris which coats teeth is known as plaque. This can harden to produce tartar or calculus.

9. Male and female gametes or sex cells have only half the number of chromosomes found in other cells. When the nuclei of male and female gametes meet and fuse, the resultant zygote has all of the genetic material it needs to produce a fully grown adult. Humans grow up with the inherited genetic characteristics of both biological parents.

10. Colds and flu are caused by viruses. Antibiotics are used to treat bacterial infections.

11. You should always seek health and safety guidance to ensure that you are not putting yourself and the children in danger from any foreseeable and avoidable risk.

12. Sketches, labelling and notes will vary. Check against text and chapter illustrations.

13. The main vertebrate classes within Phylum Chordata include fish, amphibians, reptiles, birds and mammals. The invertebrate classes within Phylum Arthropoda include arachnids, chilopods, diplopods, crustaceans and insects. Check all characteristics and classifications against the text. Annelids and molluscs are separate phyla. They each contain invertebrate organisms commonly examined in primary classrooms.

Chapter 4 – Continuity and change

1. Look at the genus and species names. The tits are in the same genus and so are very closely related, but the two crows are varieties of the same species and readily interbreed where their ranges overlap. (The hooded crow is common in North Scotland, while the carrion crow lives in South Scotland and England.)

2. New Zealand and Australia were isolated from the other continents by a fairly wide expanse of ocean and so evolution proceeded along different lines.

3. The same number in both – 46.

4. The egg has only 23 chromosomes – those from the mother only. The fertilised egg has 46 chromosomes – the mother's and the father's.

5. Their next child runs exactly the same risk of having the disease as the other two. The next child runs a 1 in 4 chance of inheriting the disease.

6. An example of a *for* argument: Genetically engineered organisms, such as the vitamin-rich rice, will have considerable benefits to poor people. The risks will be controlled by scientists who will not allow harmful food to be eaten or to cross with other organisms.

   An example of an *against* argument: Genetic engineering will mainly benefit rice companies who pay little regard to the possibility of their new organisms crossing with living things in the environment with highly unpredictable results. The food produced by genetic engineering may have consequences unforeseen by the scientists who produced it.

7. Fish developed first, followed by amphibians, reptiles, birds and mammals.

8. Humans can walk upright, have a large brain, use tools extensively, bury dead with ceremony and create art.

9. After the demise of the dinosaurs there were significantly fewer large carnivorous marine animals. The ancestors of the whales and dolphins were land-based carnivores which were able to exploit the niche left by the dinosaurs. Sea-going characteristics such as streamlined shape and top-of-head breathing holes were selected since the individuals which possessed them had an advantage over more land-adapted types in the struggle to get food. Whales and dolphins did not evolve gills because they are mammals with a highly developed form of

breathing which could easily adapt to marine existence.

10. Humans evolved like all other animals in response to evolutionary pressure. Evolution occurs by chance and often the response to a pressure in the environment is not perfect. For instance, backache is a result of imperfect spines evolving as a response to upright gait.

Chapter 5 – Ecosystems

1. Sun → grass → sheep → human
        primary   primary   secondary
        producer   consumer   consumer

    Sun → courgette → human
        primary   primary
        producer   consumer

2. There are eight food chains in the food web.

3. • Buzzard population would probably drop.
    • Caterpillar population would probably increase.
    • Availability of leaves would probably decrease, if the caterpillar population increased.
    • Other animals that eat caterpillars, forming part of other food pyramids, might increase in number, because there would be less competition for the caterpillars from blue tits.

4. Phytoplankton get their food energy from the Sun.

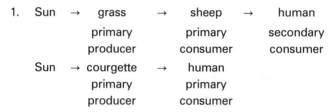

5.
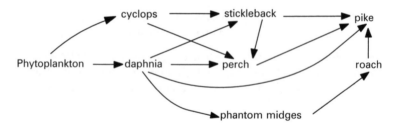

6.
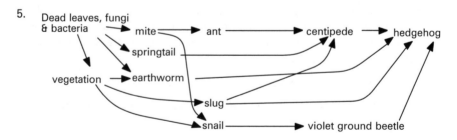

PRIMARY PRODUCER	PRIMARY CONSUMER	PRIMARY/SECONDARY CONSUMER	TERTIARY CONSUMER	QUATERNARY CONSUMER
	PREY	PREDATOR & PREY	PREDATOR & PREY	PREDATOR
DECOMPOSER	HERBIVORE	CARNIVORE	CARNIVORE	CARNIVORE

Chapter 6 – Materials

1. Both are found in the nucleus of an atom and both have an atomic mass of 1. Protons have a positive charge whereas neutrons are electrically neutral.

2. Sulphur has an atomic mass of 32. The nucleus contains 16 protons and 16 neutrons.

3. Ionic bonding occurs when electrons are removed or added to an atom; covalent bonding is when electrons are shared.

4. It isn't chemically different; it is physically different.

5. a) Each of these elements has a complete electron shell (or stable electron configuration). They are called the 'noble gases'.
   b) They are both elements with only one electron occupying the shell furthest from the nucleus.
   c) An oxygen ion would have two excess electroncs and so would be represented as O^{2-}.

   d)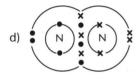

   e) MgO – ionic bonding.   f) C_2H_5OH – covalent bonding.

   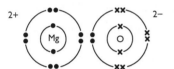   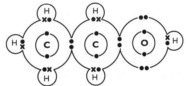

Chapter 7 – Particle theory and the conservation of mass

1. Some of the chemicals from the tea leaves dissolve in the water but the leaves themselves do not dissolve.

2. It comes from the air. Air is a mixture of gases, including oxygen.

3. It usually comes from the air. Air is a mixture of gases, including oxygen.

4. The overall chemical reaction is identical.

5. When salt is added to water, the slight negative charges in the water are able to attract the sodium chloride ions and pull them off the salt crystals, thus dissolving the salt.

6. The cap would be forced off the bottle. When water freezes it expands. This means that the same amount of water will occupy a bigger volume as a solid (ice) than as a liquid (water). If the bottle was full of liquid water, then the frozen water would need more space and so the cap would be pushed off to make the extra space.

Chapter 8 – Electricity and magentism

1. A higher resistance because it is very thin and made from a metal which does not conduct as well as copper.

2. The wire will resist the electricity and become hot and eventually melt. Thick wires should always be used on extension flexes.

3. The miles of wire underground or hanging from pylons present a substantial resistance to the flow of electricity. Removing the heat produced is difficult especially where wires are buried underground.

4. Yes. They will act like one thicker wire resulting in less resistance.

5. It is resisting electricity and converting some into heat.

6. i)   Equally bright (equal resistance)
    ii) Bulb (c) is brighter than (b) (the electricity has to work harder to get through the higher resistance bulb, therefore more energy is converted in this bulb).
    iii) Bulb (b) is brighter than (a) because Bulb (b) has less resistance so more electricity goes down this pathway than down the pathway controlled by high resistance Bulb (a).

7. The short circuit occurs when there is no resistance in one of the pathways of a parallel circuit.

8. (a) 100W bulb: 10 hours.   (b) 20W fluorescent tube: 50 hours.

9. (a) 8kW shower: 4 800 000J.   (b) Yes. The kettle will use 10 800 000J.

10. No. Tin is not magnetic. Tin cans are attracted because tin cans are actually steel cans coated with a very thin layer of tin to prevent rusting.

11. In the series circuit the resistance of both bulbs is combined and the flow is strongly resisted. There is less resistance to electrical flow when there is only one bulb (resistor) in the circuit.

12. The bulb has the higher resistance because the kettle allows more electricity to pass through it.

13. The flex leading to the kettle has to carry a greater flow of electricity. If the kettle's flex was thin, it would get hot because of the high resistance of the thin wire.

Chapter 9 – Energy

1. 3000 joules.

2. 500 × 7 200 = 3 600 000 joules.

3. No, the pan will have more heat energy. If you dropped the very hot drawing pin into a pan of warm water, there would not be enough energy transferred from the pin to the water to cause it to boil.

4. Strain energy – the springs are compressed and release their energy when you get off the bed.

5. Water. It is formed when the hydrogen in the petrol burns. On cold days, it condenses in the cool parts of the exhaust.

6. In all the cells of the body. The sugars are transported there in the bloodstream.

7. Much of it is converted to power the life processes and the heat and movement of the living things.

8. Just over 100g.

9. The tea is kept hot because (1) conduction is kept to a minimum by the thin wall of the flask, (2) convection is minimised by having two layers of glass separated by a vacuum (convection needs a gas to transfer the heat from one wall to the other), (3) radiation is minimised by silvering the glass so reflecting much of the heat back into the flask.

10. No, it would warm it, since there is an input of electrical energy extracted from the body of the fridge which is simply released into the room.

11. More than the mass of the original candle because burning the wax is a constructive process in which the oxygen is combined with the elements which make up the wax.

12. They are pulling apart the oxygen and carbon from the carbon dioxide and they are separating the oxygen from the hydrogen in the water molecules.

Chapter 10 – Forces and motion

1. A helium-filled, fairground balloon and a hot air balloon all float upwards in air.

2. The force which is the gravitational attraction between the Earth and Moon is exactly the correct size to enable the Moon to orbit the Earth without flying off into space. (If the force happened to be smaller, the Moon would travel on a wider orbit.)

3. The friction between an ice skate and the ice is reduced by having a smooth surface on the base of the skate, reducing the area of contact with the ice and using the lubricating effect of the melted ice directly under the ice skate.

4. In each drawing the ball or the brick will have a force acting vertically downwards on it which is its weight.

    In addition to this force:
    a) The ball will have a force acting upwards on it, equal to its weight, provided by the table (the reaction force).
    b) The brick will have a friction force acting up the slope and a reaction force provided by the slope acting at right angles out of the slope.
    c) The ball will have a force smaller than its weight acting upwards on it. This is a small amount of air resistance.
    d) The ball will also have a small air resistance force acting in the opposite direction to the movement. (Note: there is *no* force acting in the direction of movement once the ball has left the hand.)
    e) The ball will have an upthrust force acting upwards which is equal to its weight.
    f) The brick will have an upthrust force acting upwards which is smaller than

its weight. There will also be a small force acting upwards which is water resistance.

g) The brick will have the same upthrust force acting upwards as in question 6 (smaller than its weight). There will also be a reaction force acting upwards from the bottom of the container. The upthrust and reaction will equal the weight of the brick.

Chapter 11 – Light

1. We usually see lightning before hearing the thunder. Loud sounds made across a sports field, such as the click of a ball on a cricket bat, can be heard slightly after seeing what makes the sound. Jet aircraft often appear 'ahead' of the sound they make.

2. Some of the chemical energy in the battery will have changed to electrical energy flowing in the circuit. Some of this, in turn, will change to heat and light energy in the bulb.

3. The pupil which was covered will have adjusted to the lack of light by opening up. When you look in the mirror this pupil will appear larger than the other!

4. Make a shadow with a single light source in a place where there can be no reflected or scattered light. A room with blackened walls or a flat area of ground outside at night might be the nearest we can get to this. A shadow formed by moonlight in the desert would be an example but starlight would certainly spoil a perfect shadow.

5. The brain detects orange when mostly red and green light strikes the retina, so blue light must be absent to get the sensation of orange. Blue light is mostly absorbed by orange things.

6. The red filter would block the blue and green light. The blue would block the red and green light so no light would get through. The screen would be black if the filters were perfect.

7. Red, blue and green are the primary colours of light.

8. Light from a projector in a darkened room.
   Sunlight shines through clouds or tree tops.
   Shadows often have sharp edges.
   Laser beam seen as a thin straight line.
   Look through a thin straight tube.
   A light source can be viewed through holes made in three cards.

9. The neon striplight will give a fuzzy-edged shadow (with a penumbra).
   The point source (such as the Sun) will give a sharp-edged shadow.

10. The ball will absorb most of the red and green light from the white light falling on it and reflect or scatter most of the blue into the eye.

Chapter 12 – Sound

1. The whispering sound is made by vibrating the air in the throat so it cannot be felt by fingers on the voice box. The vocal chords do not play a part in making the sound.

2. (a) The display would show tall 'waves' packed closely together.
    (b) The sound would be quiet and of a low pitch – such as the hum you hear in a car as it travels along the road.

3. We would see a flash of light but we would not hear a sound because sound cannot travel through the vacuum in space.

4. The amplitude of a vibration is the distance the vibrating object (or molecules of air) moves from its starting point to its furthest position. A larger amplitude will make a louder sound.

5. The bottles should be arranged so that the first has the least water in and the last has the most. (This gives a descending scale of notes.)

6. Answers might include:
    - placing an ear to a desk and making a quiet sound by scratching the desk;
    - placing an ear to a large metal construction such as railings or a central heating system and making a quiet sound well away from the ear;
    - using a string telephone;
    - listening to a sound on the other side of a wall by placing a drinking glass to the wall.

7. The child is confusing cause and effect by suggesting that the sound makes the vibrations.

    The child may believe that sound will not travel through solids such as the door. The child knows, however, that the sound will be heard faintly and has explained this by suggesting that the sound travels through the gap under the door.

    The child appears to believe that sound travels by the movement of the air as though an air current is necessary.

    The child shows no sign of understanding that sound can be reflected.

    The child would need to be asked about whether the sound from the drum moves in all directions.

    It is not clear from the drawing that the child understands what a vibration is.

Chapter 13 – The Earth and beyond

1. Gravity, one of the most fundamental forces of nature. Gravity is responsible for forming galaxies and stars and probably planets throughout the Universe even today. Gravity also keeps our feet on the ground wherever we stand on the Earth.

2. The arms of spiral and barred spiral galaxies. Their appearance is similar to that produced by a spinning Catherine wheel. Astronomers have confirmed galactic rotation using more advanced scientific techniques.

3. Most analogies are productive in that they use comparisons with more familiar objects and events to make difficult concepts and processes easier to understand. Stars, of course, are not like humans and other animals or plants as they are not alive.

4. We can see planets, moons and other objects in the Solar System because they reflect light from the Sun.

5. Looking at the Sun without proper protection can damage eyes and in severe cases cause permanent blindness. You should *never* look at the Sun through binoculars or a telescope.

6. The Sun, the Moon, stars and visible planets all appear to 'rise' in the east and 'set' in the west. The most obvious conclusion to draw was that a motionless Earth sat at the centre of all astronomical activity. As detailed observational evidence and other ideas emerged over the millennia that followed, particularly during the sixteenth and seventeenth centuries with the work of Copernicus and Galileo, this view was eventually challenged.

7. The Sun, the planets, moons, asteroids, comets and meteoroids are the major components of the Solar System.

8. As the Earth rotates from west to east, people living in London would see the Sun rise first.

9. A year is 365.25 days long. This is the time taken for the Earth to complete one orbit of the Sun. But how is that known? If you could observe where the Sun appears on the horizon on any given day of the year, it would take 365.25 days for the Sun to return to that same position again, having moved once through all of the variations described.

10. A lunar year is 12 × 29 days 12 hours long. A 354-day lunar year is 11.25 days shorter than the time taken for the Earth to orbit the Sun. Keeping time with the Moon results in problems associated with fixing the time of the seasons. Certain religious festivals, including Ramadan and Easter, are linked to lunar cycles. This is why they vary in time every year.

11. This is because the plane of the Moon's orbit around the Earth is tilted at an angle of about 5° relative to the plane of the Earth's orbit around the Sun. Most of the time the Sun, the Earth and the Moon are never perfectly aligned.

12. Concept maps will vary. Check against text and chapter illustrations.

13. Pictures will vary. Check against text and chapter illustrations.

14. In June, the northern hemisphere is tilted towards the Sun and experiences summer. Days are long, the Sun's rays reach the surface of the Earth at high angles, and the Sun's heating effect is more efficient. The Sun 'rises' in the northeast, 'passes' at its highest in the sky, and 'sets' in the northwest. At full moon, the Moon would only be visible at night. Check its location against text and chapter illustrations.

Glossary

acceleration: the rate of change in velocity – measured in metres per second per second.

acid: a substance which forms hydrogen ions (H^+) when dissolved in water and has a pH of less than 7.

adaptation: a species' genetic adjustment to environmental conditions through the long-term process of natural selection.

air resistance: see **friction**.

algae: organisms including seaweeds currently classified as protoctists rather than plants on the basis of their life cycle rather than appearance.

alkali: a base that dissolves in water, has a pH of more than 7 and neutralises acids.

allotrope: giant structures made of the same element but with different structures and different properties.

amplitude: the amount of up and down movement in a wave. The size of the amplitude of a vibration will determine the loudness of the sound produced.

amps: a measure of the flow (current) of electricity.

Angiospermophyta: phylum of the plant kingdom. Includes flowering plants.

Animalia: kingdom. Complex, multicellular, eukaryotic organisms. Cells have no cellulose cell walls, chloroplasts or large, permanent vacuoles. Includes vertebrate and invertebrate groups. Humans are animals.

Annelida: phylum of the animal kingdom. Includes true segmented worms such as earthworms, leeches and ragworms.

aphelion: the point where a planet is furthest from the Sun.

Arthropoda: phylum of the animal kingdom. Includes spiders, centipedes, millipedes, woodlice and butterflies.

asteroids: small, rocky objects which orbit the Sun. Sometimes referred to as minor planets.

atom: smallest particle of an element consisting of electrons and a nucleus.

atomic mass: the sum of the number of protons and neutrons in the nucleus of any element.

atomic number: the number of protons in the nucleus of any element.

autotrophs: food producers, e.g. green plants, that use the Sun's energy to make food by photosynthesis.

balanced forces: will hold an object still or allow it to continue with a constant velocity.

base: a substance that forms hydroxyl ions (OH^-) when dissolved in water, has a pH of more than 7 and neutralises acids.

battery: a group of cells.

Big Bang: an explosion of space and time from which the Universe emerged. This is now thought to have occurred about 14 billion years ago.

black: a totally black object will reflect no light.

boiling: the rapid change of a chemical from a liquid to a gas that involves the formation of bubbles.

bond energy: the energy needed between each atom in a molecule in order to hold the atoms together.

Bryophyta: phylum of the plant kingdom. Includes common mosses, liverworts and hornworts.

buoyancy: see **upthrust**.

burning: a chemical reaction that involves oxygen and that is exothermic.

carnivore: animal (occasionally a plant) that consumes and gains energy from other consumers, i.e. other animals, either herbivore or carnivore.

cell: a container of chemicals which react together producing a flow of electrons from one terminal to the other.

cells: microscopic building blocks and basic units of life.

chemical formula: the symbolic representation of a molecule or compound.

chlorophyll: green substance within the chloroplasts of certain plant cells. Absorbs energy in light from the Sun and makes it available to carry out photosynthesis.

Chordata: phylum of the animal kingdom. Includes vertebrates and other organisms with a notochord or stiffening rod providing internal support.

chromosome: a long thin structure in the nucleus of each cell. Humans have 46 chromosomes, 23 of which are from the father and 23 of which are from the mother. Each chromosome is made from proteins and DNA.

circuit: components linked by wires which allow a flow of electricity.

Cnideria: phylum of the animal kingdom. Includes jellyfish, corals and sea anemones.

colour filter: prevents certain colours of light passing through while allowing through others.

comet: chunk of ice and dust in orbit around the Sun.

community: a group of plants and animals in a particular habitat.

compound: a substance composed of two or more elements in definite proportions by weight.

compression: the squeezing together of the molecules when vibrations are set up in a gas.

condensation: process of changing from a gas to a liquid.

conductor: a material which allows electricity to flow through it.

Coniferophyta: phylum of the plant kingdom. Includes conifers and firs.

conservation of energy: the total energy remains unchanged as it changes from one form to another.

covalent bonding: bonding between atoms where electrons are shared.

Darwin: the scientist who first described the theory of evolution in his book *The Origin of Species*.

day and night: the day and night cycle is caused by the rotation of the Earth about its axis once every 24 hours.

deceleration: the rate of decrease in velocity (some scientists say deceleration is negative acceleration).

decomposers: living things that get their energy from dead and decaying matter.

digestion: the physical and chemical breakdown of food.

dissolving: the process of making a solution where at least one of the chemicals is in a liquid state.

DNA: a chemical – deoxyribonucleic acid. It forms long double strands wound round each other in the form of a double helix. The strands are joined by chemical bonds.

dominant: in each pair of chromosomes there are genes which code for the same characteristic. If the genes programme for different effects, the one which is expressed in the characteristics of the organism, e.g. hair colour, is dominant.

Echinodermata: phylum of the animal kingdom. Includes starfish and sea urchins.

eclipse: the blocking off of light from one astronomical body by another.

ecology: the study of the relationship of living things in their natural environment. The natural environment will include living and non-living things.

ecosystem: an integrated unit of a community of living things and the physical environment in a certain area, i.e. a given community and its habitat. Studies of ecosystems will focus on interactions between living and non-living things and on the flow of energy and materials between these parts.

efficiency: the proportion of energy which is usefully converted.

elastic energy: energy in the change of shape of an object.

electricity: a flow of electrons from a negative to a positive terminal.

electromagnetic spectrum: a family of wave types which includes light, X-rays, radio waves, etc.

electrons: particles which surround the nucleus of an atom and have a negative charge.

element: substance consisting of one type of atom only.

endothermic reaction: a chemical reaction that takes energy from its surroundings because the total bond energy needed between atoms in the new chemicals is more than the total bond energy in the original reactants.

energy: the capacity to do work by moving or heating.

environment: the conditions existing in a habitat that will affect a particular organism.

eukaryote: eukaryotic organisms are composed of eukaryotic cells that have a well-defined nucleus containing each cell's genetic material or DNA.

evaporation: process in which a liquid changes to a gas.

evolution: the process by which organisms have changed over millions of years into different forms. There is a clear sequence, for instance, from dinosaurs to birds.

exothermic reaction: a chemical reaction that gives energy to its surroundings because the total bond energy needed between atoms in the new chemicals is less than the total bond energy in the original reactants.

filament: the very fine wire in a light bulb.

Filicinophyta: phylum of the plant kingdom. Includes ferns.

fission: this occurs when an unstable nucleus splits into two fragments.

food chain: the feeding relationship in a given ecosystem which illustrates the flow of energy.

food pyramid: a representation of the relative biomass in a food chain and hence, the relative proportions of energy transfer.

food web: a complex interaction of several or many food chains within an ecosystem.

fossil fuel: a fuel such as oil, coal and gas, which is stored in rocks. Energy is obtained when these fuels are burnt.

freezing: process of changing from a liquid to a solid.

frequency: the number of waves passing a given point each second, measured in hertz or megahertz. The frequency of a vibration determines the pitch of the sound produced.

friction: (including **air** or **water resistance**) the force which occurs between two substances and that tends to reduce movement or prevent potential movement.

fuel: a source of heat energy obtained from materials that burn.

Fungi: kingdom. Complex, eukaryotic organisms with thread-like hyphae rather than cells. Hyphae may contain one or more nuclei. Chitinous cell walls. No chloroplasts. Includes mushrooms, yeast and moulds.

galaxy: an assemblage of stars, nebulae and other interstellar material.

gas: a state of matter that completely fills its container and can be compressed.

gene: a section of a chromosome which codes for the production of a particular protein.

genetic: pertaining to genes or inherited characteristics.

giant structures: large numbers of atoms that are bonded together to make a new substance that is very difficult to break apart.

gravitational potential energy: energy due to the position of an object above the Earth's surface.

gravity: the force of attraction between all objects with mass.

habitat: the natural home of a group of plants and animals, that provides all (or nearly all) the needs of the inhabitants.

heat: a form of kinetic energy.

herbivore: animal that consumes, and gains energy, *only* from primary producers, i.e. plants.

heterotrophs: organisms not capable of manufacturing their own food. Humans are heterotrophs. They get their food by eating plants and other animals.

heterozygous: a person who has two different genes for the same characteristic. One of these genes will be recessive and one will be dominant.

homozygous: a person who has two identical genes for the same characteristic.

hydrogen: the lightest gas. It is the commonest element in the Universe.

hydroxide: a compound which contains a hydroxide ion (OH).

igneous rock: rock formed from magma that is solidified.

inherit: to have genetic information passed on from a preceding generation.

insulator: a material which does not allow electricity to pass through it.

ion: an atom that has lost or gained one or more electrons and as a result has a positive or negative charge.

ionic bonding: bonding between atoms where electrons are taken or given (transferred).

isotope: atoms of the same element which have the same atomic number but different atomic mass because the number of neutrons varies.

key: device which can be used to systematically classify or identify living and non-living things. Keys come in different forms.

lichen: a symbiotic association between a fungus and an alga. Lichens grow very slowly and are usually found encrusting rocks, gravestones and roofs.

light: a form of energy which affects the nerve cells on the retina of the eye. Light travels in straight lines in waves from a source.

lignification: process of impregnating and strengthening xylem cells with lignin. In woody plants like trees, fully lignified xylem forms woody tissue.

lignin: a complex chemical compound most commonly derived from wood, and an integral part of the secondary cell walls of plants and some algae.

liquid: a state of matter that takes up the shape of its container and can be compressed a little.

loudness: the volume of a sound which is a measure of the energy used to make the sounds and is determined by the amplitude of vibrations.

Lycopodophyta: phylum of the plant kingdom. Includes club mosses.

mass: the amount of matter in a substance – measured in grams or kilograms.

melting: process in which a solid changes to a liquid.

Mendel: a monk who in the late nineteenth century propounded the basics of inheritance through dominant and recessive genes. In 1865 he published an account of breeding peas on which the whole science of genetics is based.

metal: an element which can ionise by electron loss.

metamorphic rock: igneous and sedimentary rock that has been changed by heat and pressure.

meteoroids: particles of dust and rock fragments frequently derived from comets.

mineral: a naturally occurring chemical compound, often crystalline in structure.

mixture: different elements and compounds mixed together with no chemical reaction between them.

molecule: a group of two or more atoms bonded together.

Mollusca: phylum of the animal kingdom. Includes slugs and snails as well as squids and octopuses.

Monera (prokaryotae): kingdom. Usually simple, unicellular or colonial, prokaryotic organisms. Cells lack a well-defined nucleus. Includes bacteria and blue-green bacteria (formerly blue-green algae).

moons: objects which orbit planets. The Earth's moon is referred to as *the* Moon.

mutation: a sudden alteration in the genetic information which causes a difference in the appearance of the organisms. Changes which occur only in the body cells affect only the individual concerned. If the change happens in the sex cells then the change is passed on to future generations.

natural selection: the process by which individuals of the same species compete for limited breeding opportunities and food. Those which are less well suited to the environment do not obtain mates or food and so cannot pass on their genes.

nebulae: clouds of interstellar gas, ice and dust.

Nematoda: phylum of the animal kingdom. Includes free-living or parasitic roundworms.

neutron: particle found inside the nucleus of an atom with no charge and an atomic mass of 1.

nuclear energy: energy released when a heavy nucleus splits.

omnivore: animal that consumes, and gains energy from, primary producers as well as from other consumers.

organ: group of tissues that work together to carry out a particular function.

oscilloscope: an instrument which converts sound energy into a visual, graphic display on a screen.

oxide: a compound that contains oxygen.

oxygen: a gas that comprises approximately 21 per cent of the Earth's atmosphere. It is vital for life.

pH: a hydrogen ion index measuring acidity and alkalinity.

parallel circuit: an electric circuit which provides two or more pathways for electricity.

parasite: an organism that lives in or on another and which feeds on the other without giving anything in return, and often at the expense of the host's welfare.

pathogens: disease-causing micro-organisms including bacteria, viruses, fungi and protoctista.

perihelion: the point where a planet is nearest to the Sun.

phases of the Moon: how the Moon looks as it orbits the Earth. Viewed from the Earth, the Moon appears to change its shape in a cycle of phases which lasts about one month.

photosynthesis: the process by which plants manufacture their own food. Plants use carbon dioxide, water and the energy in sunlight to produce simple sugars like glucose which they can then use during cellular respiration and to produce many of the other substances they need.

pitch: the highness or lowness of a sound which is determined by the frequency of the vibrations.

planet: large spherical object which orbits the Sun.

Plantae: kingdom. Complex, multicellular, eukaryotic organisms. Most cells have cellulose cell walls, chloroplasts and a sap-filled vacuole. All green plants are capable of photosynthesis.

Platyhelminthes: phylum of the animal kingdom. Includes free-living flatworms and parasitic tapeworms and flukes.

power: energy transferred per second.

predator: a carnivore, i.e. an animal that gets its energy from consuming other animals.

pressure: when a force acts over a given area, the pressure is a measure of the force on each unit area – measured in newtons per square centimetre or pounds per square inch (as in tyre pressure).

prey: the animal food source for predators.

primary consumer: animal that consumes, and gains energy from, primary producers, i.e. plants.

primary light source: has its own supply of energy and can be seen when there is no other source of light.

primary producer: food producer, like green plants, that uses the Sun's energy to make food, by photosynthesis.

Prokaryote: prokaryotic organisms are composed of prokaryotic cells. Prokaryotic cells have no well-defined nucleus. The genetic material or DNA of prokaryotic cells may cluster or appear dispersed throughout each cell.

Protista: kingdom. Usually simple, unicellular, eukaryotic organisms. Cells may appear 'animal-like' and 'plant-like'. Includes the amoeba and the paramecium. Euglena and others are capable of photosynthesis. Algae are currently classified as protoctists.

proton: particle found inside the nucleus of an atom with a positive charge and an atomic mass of 1.

pupil: a protected hole in the eye through which light travels to the retina.

quality of sound: describes the effect of a sound's many component frequencies. This enables us to tell the difference between a note played by one instrument and the same note played on another instrument.

rarefaction: the moving apart of the molecules when vibrations are set up in a gas.

reaction force: the force created by a surface as the result of another force acting on it.

recessive: in each pair of chromosomes there are genes which code for the same characteristic. If the genes programme for different effect, the one which is not expressed in the characteristic of the organism, e.g. eye colour, is recessive.

reflection: an image formed in a reflective surface such as a mirror.

renewable energy: energy which comes from the Sun or the motion of the Moon such as winds, solar energy and tides.

resistance: a measure of the difficulty electricity has in passing through a conductor.

respiration: the oxidation of chemicals which supply energy when broken down. This happens in cells.

retina: the internal back wall of the eye containing nerve cells which are sensitive to light.

rusting: a chemical reaction that involves oxygen and results in the formation of an oxide.

salt: a compound in which all or part of its hydrogen has been replaced by a metal.

saturation: a liquid is saturated when it cannot dissolve any more solid.

seasons: the cycle of seasons is caused by the tilt of the Earth's axis relative to the plane of its orbit around the Sun. In the UK, this cycle includes spring, summer, autumn and winter.

secondary consumer: animal that consumes, and gains energy from, primary consumers, i.e. herbivore animals.

secondary light source: reflects or scatters light from a primary source, e.g. the Moon, a sheet of paper or a wall.

sedimentary rock: rock formed from deposits of weathered rock and, sometimes, the hard parts of organisms.

series circuit: a circuit where all the components are linked so that electricity has only one pathway through all of them.

shadow: the total or partial absence of light.

short circuit: a pathway in a parallel circuit which has no resistance – it allows electricity to flow unchecked.

Solar System: the Sun (our nearest star) together with its family of planets, moons, asteroids, comets and meteoroids.

solid: a state of matter that holds its own shape and cannot be compressed.

sound: a form of energy which is transmitted by waves of vibrations through a medium.

speed: the rate at which something moves from one place to another – measured in miles per hour or metres per second.

Sphenophyta: phylum of the plant kingdom. Includes horsetails.

star: normal stars are massive balls of hot, glowing gas. Stars change as their life cycle evolves.

sublimation: process of changing from a solid to a gas, without going through the liquid state.

suspension: where the small particles of a solid do not dissolve in a liquid.

temperature: the degree of hotness of something.

tissue: group of specialised cells that work together to carry out a particular function.

transverse wave: a wave in which parts move to and fro at right angles to the direction in which the wave is moving.

trophic levels: the different points or layers in a food chain or food pyramid.

unbalanced forces: will enable an object to accelerate or decelerate (including changing direction).

Universe: everything that exists: matter, radiation and space.

upthrust: the upwards acting force provided by a liquid on objects which are immersed in it.

variable resistor: a device which can be made to have a range of resistance.

velocity: the speed of something in a particular direction.

vibration: a regular backward and forward movement of a physical substance.

virus: viruses invade living cells and use them to reproduce. Viruses are pathogens and generally cause harm. The nature of viruses remains uncertain.

voltage: a measure of the energy of electrical flow.

water resistance: see **friction**.

watt: energy transferred at the rate of 1 joule per second.

wavelength: the wavelength of a wave is the distance between similar parts of successive waves. The length of a lightwave determines its colour.

weight: the force on an object due to the gravitational pull of the Earth – measured in newtons.

work: a transfer of energy as a result of a force acting through a distance; 1 joule is 1 newton moved through 1 metre.

Added to the page reference 'f' denotes a figure and 't' denotes a table.